Dance
Psychology

Dance Psychology

The Science of Dance and Dancers

Peter Lovatt

Dr Dance Presents

First Printing: 2018

ISBN 978-0-244-96056-8

Dr Dance Presents
Norfolk, UK

www.drdancepresents.com

U.S. trade bookstores and wholesalers: Please contact Dr Dance Presents see www.drdancepresents.com for details

"Dance first. Think later. It's the natural order."
- Samuel Becket.

Preface

This book is about the Psychology of Dance. Dance Psychology is the study of dance and dancers from a scientific, psychological perspective. This book is an academic textbook for students of dance, psychology or those with a general interested in the area.

Dance has always played an important part in my life. I used to love watching dancers on TV when I was a kid; those dance routines from Chitty Chitty Bang Bang still make me grin like a Cheshire cat. I don't remember when I started to dance exactly, but rhythmic movements just started coming out of me when I was about four, and the feelings associated with movement made me happy, and they still do. I earned my living as a professional dancer working in musical theatre shows and enjoyed the mental challenge of learning new movement patterns and then performing them with pizzazz.

I left the professional dance world to study Psychology. As I studied for degrees (B.Sc., M.Sc., Ph.D.) in Psychology and Computational Neuroscience, I became fascinated by how our behaviour is influenced by our genes, the people we interact with, our development, our cognitive systems and our biological make up. The study of psychology is the study of fundamental human behaviours. From spending my life in dance, I believe that dance is a fundamental human behaviour (see Chapter 1). I decided, therefore, to set up the Dance Psychology Lab and study the Psychology of Dance.

I am an academic psychologist, before that I was a professional dancer, and before that I was an illiterate schoolboy; these are the three voices I heard in my head as I wrote this book. I hope you enjoy my exploration of Dance Psychology

Contents

INTRODUCTION

What is Dance Psychology?

Dance Psychology is the study of dance and dancers from a scientific, psychological perspective.

I set up the Dance Psychology Lab, at the University of Hertfordshire, in 2008 because I become aware of a growing body of academic literature which revealed dance and dancers through the lens of psychology. These individual pieces of research were being carried out in research labs around the world. For example, a project carried out at Rutgers University in the U.S.A. found that the way people dance, especially men, is related to their genetic make-up. What William Brown et al. (2005) found seemed remarkable to me. They showed a relationship between a person's genetic make-up and the way that person danced. Brown et al. reported that the dancing of those people who were more physically symmetrical was rated as better quality by groups of observers than the dancing of people who were less physically symmetrical. This, together with similar research, which looked at the relationship between dancing and hormones carried out in Germany, by Bernhard Fink (see Fink et al., 2007), and in the UK by Nick Neave (see Hugill et al. 2010) made me look at the act of dancing, and our love for doing it, in a completely new way. But it wasn't just about hormones.

Another piece of research, this time from a laboratory in Italy (see Di Blassio et al., 2009) reported that the characteristics of Caribbean dance are useful for the promotion of better health. It is clear that physical inactivity is associated with higher rates of chronic disease and premature death, but now there was a growing body of research evidence which suggested that dancing may have non-trivial positive effects on our health, and it might even help us to live a longer life. However, not everything was rosy in the garden of dance. Frank Bakker (see Bakker, 1988) carried out a study of young dancers in the Netherlands and reported that dancers had lower self-esteem than non-dancers. This leaves us with a conundrum; how is it possible for dancing to be both good for our health and at the same time bad for our self-esteem?

One other piece of research caught my eye as I was setting up the Dance Psychology Lab. This research was carried out by an international team from the USA, Canada and the UK, and was led by Steven Brown (see Brown et al. 2006). This project, which involved people dancing while in a brain scanner, found that certain areas of the brain become active when you dance, and in an article published in the academic journal *Cerebral Cortex* called *The Neural Basis of Human Dance* the authors claimed to have identified the interacting network of brain areas active during different aspects of dancing, and moving to music.

I found all of these papers fascinating. However, as a Psychologist and a scientist, these research findings left me with more questions than answers. Being a scientist can sometimes mean being sceptical, it certainly means asking difficult questions, interrogating the data and not necessarily taking the evidence presented at face value. This is essential, because it is only by approaching science from such a perspective that we can make significant advances in our knowledge and understanding of the world. This is also the case for developing a greater understanding of Dance Psychology.

Therefore, as a sceptical, scientific Dance Psychologist I was left asking, if dancing is related to our hormonal or genetic make-up, how exactly do our hormones make us move, does the way we move and dance change as our hormone levels change? How many sessions of recreational Caribbean dance are necessary to improve our health? Does everyone who dances get the same benefit, or do some people benefit more than others? What other forms of dance give us the same health benefits? Is Caribbean dance better for our health than Balinese dance? What sort of dancers (e.g. ballet, burlesque, ballroom) have lower self-esteem than people who don't dance, and what is it about dancing that causes the reduction in a person's self-esteem, is it something to do with dance environment, the teachers, or the characteristics of the dancers themselves? Is it really possible to dance in a brain scanner, while keeping your head and most of your body perfectly still? I set up the Dance Psychology Lab so that, with others, I could work to find answers to questions like these.

Understanding more of the detail about these Dance Psychology topics is important because if dancing is associated with certain positive or negative outcomes, we must be certain of our findings and know the limits of their application. For example, let us assume that there is some evidence that dancing makes people healthier and live longer (as there is). Now, one of the implications of this finding might be that dance sessions could be offered to everyone, paid for by the state, to enhance the health of the nation. Of course, before any government decided to pay for such a scheme it would need to have strong, compelling evidence about the relationship between dance and health; as they have, for example, about the relationship between medicines and health.

I am often asked what Dance Psychology is and what methods Dance Psychologists use to answer questions. Dance Psychology is the study of dance from a psychological perspective. Psychology is about how people think, feel and behave. It's a broad subject and in the next few pages I will set out the different areas of psychology and the types of questions that psychologists ask. Psychology is a science and most academic (or university) psychologists are scientists. Dance, in its broadest sense, is about movement. It can be about moving on your own or with other people, it can be about moving in set patterns or moving in unplanned ways, it can be about moving to a beat or in silence. Dancing can mean being free or constrained. Dance can be an art form, or a sport, and some dancers are highly trained with expertise in one or more techniques of dance. For some, dance is a vocation.

Dance can be a recreational activity, a hobby or something that is done informally at parties, weddings or in the privacy of your own home. Dancers are people who dance. A prima ballerina performing the lead role in Swan Lake at the Royal Opera House is a dancer. So too is the middle-aged man dancing to ABBA in his kitchen. You may not consider yourself to be a dancer, but if you've ever put the radio on and danced around your bedroom, then you're a dancer. In some sense, we are all dancers. In Chapter 1 I address the question of whether we are born to dance. We know, for instance, that babies will move rhythmically in response to rhythmic stimuli (e.g. music) and so, in at least that sense, we are all dancers; whether or not we have ever had a boogie in the bedroom or attended a dance class.

Psychology is about human behaviour. It's about how we think, feel, form relationships, and, basically, how we function in every way; from our brain to our behaviour. When we study the psychology of people we often break it down into these different areas.

Developmental Psychology

Developmental Psychology is concerned with studying the changes which occur in people across their lifespan. Developmental psychologists address questions such as, how do children develop language skills? and How do we cope with our changing roles throughout life, for example, on becoming a mother, or changing our job as we get older?

When we look at dance from a developmental psychology perspective we are able to address questions such as, are babies born with the innate ability to dance? Why do people stop dancing when they get to a certain age? And does dancing serve different functions for us as we move from the beginning to the end of our lives?

Social Psychology

Social Psychology is concerned with understanding people and the relationships they have with other people, groups and larger institutions. Social psychologists and personality theorists address questions such as, how is personality defined and measured? and why do some groups of people have different personalities to other groups of people? Is it their personalities which draw them together or do people's personalities change as a consequence of associating with certain other people?

When we look at dance from a social psychology perspective we are able to address questions such as, what is it about dance that draws people together? Why do different groups of people dance in different ways? and how does social dancing help to bond people together?

Biological Psychology

Biological Psychology is concerned with explaining human behaviour in terms of underlying physiological and evolutionary processes. Biological psychologists address questions such as, what physiological processes can be called upon to explain behaviour? and what are the genetic-evolutionary processes which predispose some people to certain behavioural characteristics?

When we look at dance from a biological psychology perspective we are able to address questions such as why is dance good for people with neurodegenerative disorders, such as Parkinson's disease? Why do we express our hormonal make up through the way we dance? and why does dance sometimes send us in to a trance-like state?"

Cognitive Psychology

Cognitive Psychology is concerned with understanding the mechanisms and processes involved in perception and thinking. Cognitive psychologists address questions such as, how do we recognise, interpret and remember what we see and hear? and how does our emotional state affect our performance on certain tasks?

When we look at dance from a cognitive psychology perspective we are able to address questions such as, how do we communicate through dance? How do we recognise emotions as they are expressed through body movement? and how can moving help us think differently?

Health, Counselling and Clinical Psychology

Health Psychology, Counselling Psychology and Clinical Psychology are another group of sub-disciplines, which is also relevant to Dance Psychology. These are concerned with studying, understanding and treating psychological disorders, such as depression, social phobias and eating disorders. Psychologists in this group address questions such as, what are the determinants of such psychological disorders? and what can be done to help people overcome, or come to terms with, such psychological disorders?

When we look at dance from a health or clinical psychology perspective we are able to address questions such as why do dancers have lower self-esteem than people who do not dance? How can dance be used to improve people's physical health? and why do people feel great after they've been dancing?

Psychodynamic Theories

Psychodynamic theories run parallel to these sub-disciplines, they might be based on the work of Sigmund Freud and are concerned with understanding the interplay of mental forces, particularly in relation to two basic ideas, which are that people might be unconscious of their motives, and that defence mechanisms operate to prevent unacceptable or anxiety-provoking motives from entering consciousness. Psychodynamic practitioners often work in a therapeutic environment with clients. Despite the fact that I started my journey in psychology by reading the works of Freud and Jung I do not carry out work in this field.

Research Methods

Dance Psychologists use a range of scientific research methods to collect, analyse and interpret data. For example, they might collect data under carefully controlled experimental conditions, through questionnaires, using focus groups or through one-to-one interviews. One of the broadest distinctions between these methods is whether the data are subjected to quantitative or qualitative treatment methods.

Quantitative methods
Quantitative treatment methods involve recording, for example, exactly how often something happens, or how much change occurs over a given period of time or how tall everyone is in a certain sample.

When we look at dance from a quantitative perspective we are able to measure, precisely, which type of dance has the best effect on a person's balance, or we can record much a woman's hips swing when she's at the fertile stage of her monthly cycle.

Qualitative methods
Qualitative treatment methods involve, for example, describing how it felt when something happened, or giving an unconstrained opinion on a piece of artwork. When we look at dance from a qualitative perspective we are able to get answers to questions such as, why do you dance? or how does dancing make you feel? and we can learn, in detail, how people feel, from a subjective perspective, when they dance.

Within the Dance Psychology Lab, and in the world of Dance Psychology generally, we have used these different psychological approaches to study dance and dancers and it is this research, and research drawn from many other scientific and dance-related fields, which you will read about in this book.

Is Dance Psychology the same as Dance/Movement Therapy?

No. Dance/Movement Therapy is defined, according to the American Dance Therapy Association website, as "the psychotherapeutic use of movement to promote emotional, social, cognitive and physical integration of the individual, for the purpose of improving health and well-being" (retrieved from www.adta.org 24th January 2018). While Dance Psychology is concerned with understanding the consequences of dance on the emotional, social, cognitive and physical aspects of a person, and is interested in understanding the use of dance for improving health and well-being, the focus of Dance Psychology is different to Dance/Movement Therapy, in as much as Dance Psychology is primarily concerned with understanding dance and dancers from a scientific, psychological perspective, and it is not based on principles of psychotherapeutic theory or practice.

CHAPTER 1

Are we Born to Dance?

Dance is a pervasive activity. The young, the old and middle-aged dance. Men and women from all around the world get down and boogie and, according to historians, there is evidence from cave paintings that we have been dancing for tens of thousands of years. But why do we dance? Is it a tradition, something we have learnt to do, that has been passed down from one generation to the next, or is dancing deeply rooted in our genes? Are we born to dance? In this chapter I look at research evidence from studies of very young babies and children in search of clues to help us decide whether dancing is an innate, biologically-driven, activity or whether we learn to dance because of our environmental influences. I explore three main questions:

1. Are babies born with the ability to detect a rhythm?

2. Do babies naturally make rhythmic movements when they hear a rhythmic sound?

3. Are babies' preferences for certain sound patterns influenced by how they have moved their body?

The Dance Nature or Nurture Debate

Before we address these questions, I want you to think about who you are, about your hobbies and habits. Make a mental list of the things you do, the way you speak and the types of friends you have. As you think about each of these things try to decide whether the things you do are innate and biologically driven, or whether they are learnt and socially conditioned. For example, let's assume that you have the ability to speak, communicate and understand what other people are saying, and let's assume that you speak English. Your ability to speak and understand other people might be considered an innate, biologically driven, function. No one teaches you how to speak and communicate, it is something that virtually everyone in the world can do naturally (it's like walking). On the other hand, the language we speak (for example English) is learnt socially. We learn to speak and understand the structure of a particular language based on where in the world we are born and the language we hear as we grow up. If you're born in the United Kingdom to an English-speaking family you'll probably learn to speak English, whereas if you're born in China to a Mandarin-speaking family then you'll probably learn to speak Mandarin. (For further reading on the innate versus learnt nature of language see Harley, 2013).

What about the other things on your mental list, which of those do you think is either innate or socially learnt?

One of the central questions in psychology concerns the extent to which our behaviour, perceptions and preferences are due to either hereditary or environmental influences. If we are born with a predisposition to something, such as the ability to walk and talk, then we might assume that the functions of walking and talking

22

are influenced by hereditary factors, such as our genetic make-up that we inherit from our parents. If, on the other hand, we develop a habit or a set of behaviours, like speaking with a certain accent, then we might assume that we have been influenced by environmental factors. This is known as the Nature-Nurture debate, where nature refers to heredity or a genetic predisposition to behave in certain ways, and nurture refers to learning from experience and the environment. Most of our behaviour can be thought of in terms of being influenced by both nature and nurture and one area that psychologists are interested in is understanding the relative contribution of both our genes and the environment in which we live in influencing different aspects of our behaviour.

The question I want to discuss in this chapter is whether dancing is a natural, nature driven activity or whether it is a nurtured activity. In other words, are we born to dance or is dancing something we learn to do based on our environment? When we first look at this question it seems obvious that dancing is something we learn. If, for example, you think about the superstars of classical ballet they have certainly been taught a particular technique of dance and there are dance schools, teaching people to dance, in almost every town and city. There is no doubt therefore that people are taught to dance. However, the question is then, is all this teaching necessary for people to dance? Or, put another way, can people dance without any training? If people can dance without training then it might suggest that dancing is an innate activity and if they cannot dance without training then it suggests that dancing is a purely learnt activity.

A Test of Innateness

How do we know if behaviour is innate? If something is an innate, instinctual behaviour then it needs to pass a test. There are three elements to the test of innateness: Inclination, Automaticity, and Universality.

Inclination

There needs to be a natural tendency for a living organism to display a particular behaviour. To pass this test people need to show a natural urge to dance.

Automaticity

People need to be able to behave in a particular way without the need for instruction. To pass this part of the innateness test people need to show that they can dance without being taught.

Universality

People from all parts of the world need to display similar behaviour. In other words, people from all parts of the world need to display dance-based behaviour for dancing to be considered innate.

Academic research, which provides evidence for the innateness of dance, comes from various disciplines. The universality of dance, both in time (history) and place (location) has been well documented by anthropologists, who suggest that dance-like behaviour dates back at least tens of thousands of years and has been evidenced in cave paintings and in religious rituals from otherwise disparate communities (see, for example, Farnell, 1999; Kaeppler, 1978; Reed, 1998). The inclination and the automaticity of dance has been shown in a series of academic research studies, which have tested adults (Janata, Tomic & Haberman, 2012), very young children (Zentner & Eerola, 2010) and babies, some as young as two days old (Winkler, Haden, Ladinig, Sziller & Honing, 2009). In this chapter I will describe the research findings that can be associated with inclination and automaticity.

Chicks and Goslings

Researchers who are interested in the relative contribution of nature and nurture to learning typically test very young people or animals. By testing the very young, scientists are able to determine what the baby has come into the world already equipped to do, without being taught. For example, some of the classic studies in psychology on early learning have come from research on newly hatched chicks and goslings. Chicks and goslings belong to a group of birds known as precocial. Precocial birds are those that are able to walk as soon as they hatch. Because they can walk as soon as they hatch precocial birds are able to follow their mothers. The birds are born with an innate instinct to follow their mother; however, they have to learn who their mother is. In studies reported by Spalding (1873/1953) and Lorenz (1935) it was found that young chicks and goslings would follow Spalding or Lorenz respectively if these were the first moving creatures the hatchlings saw. Lorenz famously called this process imprinting as he showed that goslings would follow him, instead of their mother, if they encountered him first. Once imprinted the learning was, apparently, irreversible.

There is therefore a tradition of examining very young creatures to understand what they come into the world knowing and doing instinctively and what they can learn in the first few hours of life.

This tradition has also been used on very young babies. But how can we test whether new-born babies arrive with the ability to dance?

Defining the Indefinable (Dance)

There are many different definitions of dancing. For the sake of this chapter I will define dancing as an automatic physical response either to other people's movement or to a heard or felt rhythm. To detect the presence of this type of dancing we need to be able to do three things. First: we need to be able to detect the beat, to either hear or feel a rhythm. For this to happen there needs to be some kind of detection of that rhythm or beat. The ability to perceive a beat helps people synchronise their movements together, so this might be synchronising across people or it might be synchronising your own sense of movement with that rhythm – so detecting the beats in music is really important, it is the first part, the first thing you need for defining what dancing is.

Second: we need to be able to move spontaneously to a beat or rhythm. If it is the case that dancing is an innate ability then we should be able to innately feel the groove and then allow our bodies to move in response to that rhythm. Third: we need to be able to learn through body movement, such that the movement of the body influences how we perceive the world. This provides a function for body movement. What this means is that while you are moving your body, something else is happening at a perceptual level. So, the movement in the body itself has some part to play in your perception of what is going on around you.

Let's start to address our three research questions.

Question 1: Are babies born with the ability to detect a rhythm?

A research team from Hungary and Holland set up a study to see if young babies perceive beat and rhythm. Of course, when scientists are looking at this question from a nature/nurture perspective they want to get hold of test participants who are very young, so that they test people who haven't been influenced too much by learning. Winkler et al. (2009) tested a group of 14 babies, that were either two or three days old. All of the babies were born at between 37 and 40 weeks gestation, and as such were considered to have undergone a full-term pregnancy.

Winkler et al. played music to the two to three day old babies. The rhythmic music consisted of a snare drum, a bass and a high-hat symbol and it had a certain predictability to it, such that once you started to hear the music you could, eventually, predict the next beat in the pattern. Using this rhythmic regularity Winkler et al. wanted to find out whether the babies themselves could pick up on the regularities within the musical structure. They argued that if the babies could do this at two or three days of age then beat perception must be innate. If, on the other hand, babies are not able to perceive a rhythm, which is something we know adults can do, then it would suggest that beat perception is learnt from the environment (a nurtured ability) and not innate.

Now, of course, Winkler et al. had a problem because you can't measure the behavioural response to music in babies very easily. You cannot ask a baby to tap along in time to rhythm or ask them when two rhythms are the same or different. So, what they did was to take a more direct measure, and as such they measured the electrical brain responses to sounds. They measured something called Event Related Brain Potentials or ERP. An ERP is a measured brain response to stimuli. In this case Winkler et al.

wanted to measure the brain's response to musical rhythms. ERP's are measured by picking up electrical signals from different parts of the brain. To do this, scientists get participants to wear a cap, something like a swimming cap, which has lots of electrodes on the inside, which touch the scalp. These electrodes pick up changes in electrical activity in the brain as people are presented with different stimuli. By putting an ERP cap on babies the scientists were able to record ERP's even when the babies were asleep.

As well as measuring a brain's response to some particular stimuli, scientists can also measure what the brain is expecting to encounter. When the brain encounters something that it doesn't expect, something that deviates from expectation, then there are a unique set of electrical patterns that can be measured. For example, when a sound pattern doesn't match what someone is expecting to hear it triggers a particular brain response. This response is called *mismatch negativity* and this can be measured using ERPs.

Winkler et al. played babies the sound of the snare drum and the bass and the high-hat symbol and then they manipulated some of the signal, such that they took out some of the beats. Let's have some examples.

Imagine, or clap, a steady, evenly paced beat that goes:

one, two, three, four,
one, two, three, four,
one, two, three, four,
one, two, three, four

Now clap it again with a stronger clap on the first beat in each set of four, so it goes:

ONE, two, three, four,
ONE, two, three, four,
ONE, two, three, four,
ONE, two, three, four

After a while you expect to hear a stronger clap on "ONE". What Winkler et al. did, rather simply and ingeniously, was to take away that strong beat from some of the signal, so it would go something like:

ONE, two, three, four,
ONE, two, three, four,
ONE, two, three, four,
one, two, three, four

You'll notice that the fourth strong beat is missing. Winkler et al. wanted to know whether the babies brain would respond to an unexpected missing beat. If a new born has a sense of rhythm then they should expect to hear beats which are predicted by the preceding rhythm and if one of these beats is missing then it will trigger the mismatch negativity response that can be picked up by the ERP.

Winkler et al. found that new born infants could detect occasional omissions of the first strong beat of a rhythmic pattern. In other words, when a beat that was predicted by the preceding rhythm was missing the babies noticed. They concluded from this that neonates form a detailed representation of a base pattern at just two days of age. It is the formation of this mental representation of a base pattern that allows them to sense the beat.

So, in answer to Question 1: Are babies born with the ability to detect a rhythm? The evidence suggests that the capability to detect a beat in rhythm, or in rhythmic sound, is already functioning at birth, and therefore beat perception appears to be innate.

Question 2: Do babies naturally make rhythmic movements when they hear a rhythmic sound?

The desire, urge or need, to move some part of our body in relation to some aspect of sound is called groove. When people feel the groove, they feel compelled to move. This derives from a process called sensorimotor coupling. The idea is that our senses either pick up the sound, or our body feels a beat, and this triggers a motor response in some part of our body (see Madison, 2006).

Two studies have shown that adults feel the groove under different conditions. Janata et al. (2012) have shown that different types of music elicit different levels of groove. They played 148 musical excerpts from different musical genres (folk, rock, jazz and R&B/soul) and asked people to rate the degree to which they feel the musical excerpts "grooves". What they found was that R&B/soul music was rated as groovier than folk, rock and Jazz, and that, in general, faster music was rated as more groovy than slower music. Of their 148 excerpts they found that Superstition, a track released by Stevie Wonder in 1972, received the highest groove rating. Superstition is a funky soul standard. Its multi-layered opening bars are characterised by orchestral and vocal syncopation and, it seems, is impossible to listen to without wanting to get down. When a rhythm is syncopated it means that for some beats there is a temporary displacement of the strong beat, such that although you are expecting to hear a beat it doesn't always appear when you might be expecting it. Syncopation means accenting the unaccented beat, or putting a stress on a weak beat.

Another example of a groovy piece of music, which Janata et al. didn't include in their study but which I'm certain would produce a strong urge to move due to its orchestration and syncopation, is a 1962 tune from Quincy Jones called Soul Bossa Nova. Have a listen, and try to keep still.

The syncopated aspect of both Soul Bossa Nova and Superstition may be one of the main drivers for eliciting a feeling of groove in adults. A study by Witek, Clarke, Wallentin, Kringelbach & Vuust (2014) found that feelings of groove varied as a function of the amount of syncopation in music. They report that medium degrees of syncopation elicit the greatest desire to move, and also the most pleasure. If there is too little, or too much, syncopation then this leads to a reduction in both desire to move and pleasure. It seems, therefore, that adults show sensorimotor coupling and feel more or less compelled to move to different pieces of music, such that higher compulsion to move is seen for faster rather than for slower music, soul/R&B over folk, rock and jazz, and medium level syncopated music. The question we need to ask now is whether human adults have learnt to feel the groove by their exposure to music during their lifetime or whether humans are born with groovy genes and the desire to move to music.

Now of course we want to find out whether babies or young children have the groove, and going back to our question about the nature/nurture debate we need to examine the psychology of the groove in very young children.

A study by Zentner and Eerolla (2010) helps us. They tested 120 babies aged between five and 24 months. They were interested in the question of whether non-verbal infants engage actively and spontaneously in rhythmic behaviour to music. In other words, they wanted to know whether babies spontaneously use rhythmic movements when they hear music. They compared baby's movement-based responses to rhythmic music and rhythmic speech. They were also interested in whether the babies enjoyed the experience of moving to rhythmic music.

Zentner & Eerola (2010) sat the babies on their mother's laps and played either music or speech through loudspeakers. As the babies sat and listened they were video recorded and these recordings were analysed for signs of rhythmic movements. A rhythmic movement was defined as a movement that is repeated three times at short intervals. It might be a movement of the arm, leg, whole body or torso.

The research team found that babies did make rhythmic movements in response to the rhythmic stimuli, and they found that babies made more rhythmic movements to metrically regular music than to rhythmic speech. While babies listened to music they made rhythmic movements that lasted, on average, between 6 and 8 seconds, whereas while they listened to speech they made rhythmic movements lasting less than 2 seconds. They also found a positive correlation between rhythmic movement and the amount of smiling that the babies did, such that the more the babies moved the more they smiled. Of course, we don't know which way round it is – we don't know which caused what, whether it was the case that the babies movement caused them to smile more or whether it was the smiling and the pleasure that made them move. Nevertheless, there is a clear link between music, movement and smiling in very young children.

These findings are discussed in terms of a social context and a neurobiological context. In terms of the social context of dance, it is suggested there is a social origin to movement coordination with music, perhaps such that as babies (and adults) move in response to music they also move in response to each other's rhythmic movements and this helps them to form a social bond. Kirschner & Tomasello (2009) suggest that the behaviour of children as young as two and a half years of age may be driven by a specific human motivation to synchronise movements with other people during joint rhythmic activity, like listening and moving to music. In this regard moving may unite people socially.

Another interesting, and relevant, study concerning the social basis of dance suggests that unified, synchronised movement may also promote pro-social behaviour in 14-month-old babies. Cirelli, Einarson & Trainor (2014) had carers bounce with babies. The carers either bounced in synchrony with the babies or out of synchrony. Later they tested the baby's pro-social behaviour by having the carer drop something and observing whether the baby would help pick it up. They report that those babies bounced in synchrony with their carers were more likely to help pick up the object than those babies bounced in an asynchronous way. Cirelli et al. suggest that motor synchrony with other people may promote the early development of altruistic behaviour. Returning briefly to the study of Zentner & Eerola (2009) lets consider the social element of dance in that study. It is clear to see that there was minimal social interaction as the babies sat on their mother's laps. We cannot tell how much movement the mothers were making themselves while they heard the music. Their own, very subtle movements in time with the music may have influenced the movements of their babies and this may also have influenced the amount the babies smiled.

In terms of the underlying neurobiological mechanisms that support the ability to move in response to music, Zentner and Eerola make reference to several brain regions that have been shown to be important in the entrainment of movement to rhythm in adults. These areas (e.g. the basal ganglia, cerebellum, vestibular system) are thought to operate in an automatic way from birth, suggesting that moving in response to a beat or rhythm may be innate, and as such does not require learning.

Although babies make rhythmic movements in response to rhythmic stimuli they do not appear to synchronise their movements to a beat, especially when they dance on their own. There have been several studies which have looked at when young children start to synchronise their movements to a rhythm and the age at which they start to do this is about four, although there is some suggestion that it might start as young as two and a half. If you would like to follow up on this please refer to the work of Provasi and Bobin-Begue (2003) and Eerola, Luke and Toiviainen (2006).

Question 3: Can babies learn rhythms from body movement?
The third question I want to look at in this section addresses the question: Do babies have the capacity to learn through movement? So, in other words, if babies move their body in a particular way will that alter how they perceive their musical environment and the world around them? This is a tricky question but in a great, and very simple, experiment (and the greatest experiments are always the simplest) Phillips-Silver and Trainor (2005) tested the hypothesis that movement influences the auditory encoding of rhythmic patterns in babies.

Phillips-Silver and Trainor took 16 healthy seven-month old babies and played them an ambiguous rhythmic sound pattern for two minutes. The sound had no accented beats, but it did have a regular rhythm. The babies sat on their parent's knees while they listed to this sound pattern. The parents played a crucial role in this study as half of them had to bounce their baby on their knee on every second beat and the other half had to bounce their baby on every third beat. So, while the babies heard exactly the same auditory stimuli their movement patterns were different. This was called the training phase of the study.

The next part of the study was a testing phase. In the testing phase the babies were played a similar piece of music to the training phase. However, this time accents had been added to the soundtrack. In one version an accent was added to every second beat and in another version accents were added to every third beat. This structure therefore complimented the different patterns of bouncing used by the parents in the first part of the study. The babies were then given a looking time preference test. This is quite clever. A flashing light attracts the attention of the baby and when the baby looks at the light a piece of music starts to play. When the baby turns their head away from light the music stops. The researchers played twenty pieces of music to each baby, such that on half the trials the baby heard the music with the accent on every second beat and the other half of the time they heard the music with the accent on every third beat. Phillips-Silver and Trainor found that the babies spent more time looking at the light (and listening to the accompanying music) of the pieces of music that had an accented pattern that matched the bouncing pattern they'd felt earlier. So, if their mother had bounced them on every second beat they later preferred the music with an accent added on the second beat and if their mother had bounced them on every third beat they showed a preference for music with every third beat accented.

This suggests that babies have the ability to feel rhythm through their body and then, just by feeling that rhythm, they prefer to listen to music which corresponds to that structure. Therefore, there is something about the embodiment of rhythm, moving your body, which changes your preference for certain rhythms. What this suggests is that the movement of an infant's own body is critical for the multisensory effect observed in these studies.

Phillips-Silver and Trainor suggest that there might be a neurological basis for this finding, which involves the vestibular, and perhaps the proprioceptive systems. "The early development of the vestibular system, and infant delight at vestibular stimulation when bounced to a play song or rocked to a lullaby, suggests that we are observing a strong, early vestibular-auditory interaction that is critical for the development of human musical behaviour." They conclude by suggesting their findings "…provide evidence that the experience of body movement plays an important role in musical rhythm perception."

This suggests that the way we move when we are with other people has an impact on our musical preferences. Although this study was carried out with babies the findings might have implications for how our adult social interactions also impact on the rhythms and musical genres we like. I wonder, for example, whether people who live in densely packed cities have different musical preferences to those who live in the countryside, and whether this might be mediated by the way our environment moves us. This brings us back to babies, and to the environment in which babies are carried before they are born.

All of the studies we've looked at so far suggest that children show an innate physical response to music and rhythm. When experimental studies use neonates as participants, particularly when the babies are just a few days old, we might assume that the baby has had very little time to develop as a consequence of nurture. But we must remember that even the two-day old baby has had several months of interaction with its mother. Such antenatal interaction is very likely to have influenced the developing foetus and so behaviours displayed at birth might not be solely due to a genetic predisposition.

This is illustrated in an interesting study by Bellieni, Cordelli, Bagnoli & Buonocore (2004) who report that the babies born to mothers who danced during their pregnancy needed to be rocked to sleep more in their first year of life than the babies who were born of mothers who didn't dance in their pregnancy. In addition, they also found that babies of dancing pregnant-mums were more likely to play musical instruments as older children and teenagers than the babies of none dancing mums.

This study highlights the fact that motherly nurturing of a baby can start several months before a baby is born and a baby's postnatal behaviour and preferences might have been influenced by their environment as well as their genetic make-up. As such, it might also be the case that a mother's behaviour, when her children are in the womb, might have an impact on the child's perception for music, perception for rhythm and the likelihood that they dance.

Summary

Can we say that babies are definitely born to dance and that it is due to nature? Well there seems to be some evidence to support this idea but we cannot yet rule out the conclusion that nurture might also play a very important role in musical and dance development too, not least during prenatal development.

CHAPTER 2

Moving and Thinking

When I was at school we were told to sit still and listen. I assume the teachers thought that if we fidgeted in our seat, or moved our legs about, it would distract us from what we were meant to be learning. I fidgeted a lot; the teacher told me countless times to sit still and listen. I got bored sitting still, and didn't learn very much.

Before my driving test in 1982 I put on Freebird by Lynyrd Skynyrd, waited five minutes, and then I danced myself crazy for the next four minutes. The music was very loud and the sitting room in my parent's house rocked. Freebird is a heavy rock anthem that starts quietly and builds to an amazing, head-banging, sweaty, exhausting crescendo. My dancing matched the build-up and became wild. As I improvised I wound up the tension in different parts of my body and then I exploded, releasing energy as I shook my head and limbs. It felt great to be alive.

It was clear to me at a very young age that sitting still made me miserable and bored and bouncing around the room made me feel alive, not just physically but mentally too. We know from the scientific literature that dancing can improve a person's mood and there is now a growing body of experimental evidence which suggests that dancing can change the way people think, solve problems and take risks. It seems that dancing around my parent's sitting room may have woken up my mind, and helped me pass my driving test.

Moving

Dance is a form of physical exercise and there is a substantial body of experimental evidence which suggests a link between physical exercise and creative thinking (for a meta-analysis see Chang, Labban, Gapin & Etnier, 2012). The effects of different forms of physical exercise, such as jogging, swimming, fast walking, riding an exercise bike and climbing the stairs, have been shown to have small positive effects on a range of cognitive processes, such as information processing, reaction times, attention, decision-making and memory. For example, Blanchette, Ramocki, O'del & Casey (2005) examined the effect of physiologically arousing exercise on a creativity task. They found that people were more creative following a short bout of exercise, where heart rate was raised to about 140 bpm, than following a period of rest. They also showed that enhanced creativity lasted for at least two hours after the end of exercise. They argue that physiological arousal enhances creativity but they caution against too much exercise, suggesting that fatigue may mitigate the positive effects of exercise on creativity. So, if you want to enhance creativity through exercise its wise to get yourself worked up, but not too worked up.

If exercise needs to be physiologically arousing to enhance thinking and creativity then what role can dance play in giving people a mental boost? A number of studies have looked at the effect of different types of dance on creativity, thinking and problem solving. In a series of studies from the 1980's Joan Gondola and her colleagues (Gondola, 1986; 1987; Gondola &

Tuckman, 1985) published the results of several studies which show that exercise and aerobic dance can both enhance creative thinking. Building on these findings, Steinberg, Sykes, Moss, Lowery, LeBoutillier & Dewey (1997) examined creativity after participants took part in either aerobic exercise or aerobic dance sessions, which lasted about 25 minutes, or after they watched a documentary about the rock formation of the Lake District.

Creativity scores were higher for those people who were in the aerobic sessions than for those who had watched the rock formation video. The higher creativity scores were shown for creative flexibility. Creative flexibility refers to how much variety a person can think of during a problem-solving task and it provides an indication of the dynamic breadth of someone's thoughts. Let us imagine you're locked out of your house and there's no-one inside who can open the door for you. Someone scoring low on creative flexibility might think about trying to open a window in order to get in to the house. All they might think about is trying to open a window, even if that proves to be impossible. However, someone scoring high on creative flexibility might also think about trying to open a window but they will also think about other ways of getting in to the house, for example, by calling a neighbour who holds a key, or finding someone with a ladder to gain access through the chimney, or gaining access through a neighbour's loft space or using a piece of wire to pick the lock. The breadth of ideas doesn't provide a guarantee of success but it does demonstrate a greater degree of generating creative possibilities. So, according to Steinberg et al., people who take part in 25 minutes of aerobic dance were found to be more creative than those who watched a video.

In addition, Steinberg et al. also report that aerobic dance is marginally more effective for improving creativity than aerobic exercise. They suggest that the nature of the exercise might have an important impact on the outcome of creativity measures. Rather than the exercise only being measured in terms of how physiologically arousing it might be (as was discussed by Blanchette et al., 2005) Steinberg et al. raise the possibility that there might be something in the freedom of the movements used in aerobic dance that also led to enhancements in creative flexibility and perhaps free movements, as opposed to a regimented aerobic exercise regime, are one of the elements responsible for unlocking creative potential.

Thinking

Support for the view that freedom of movement facilitates creative thinking comes from two experimental studies (Campion & Levita, 2014; Leung, Kim, Polman, Ong, Qiu, Goncalo & Sanchez-Burks, 2012). Leung et al. were interested in the idea of embodied-metaphors and how they might influence creativity. Embodying a metaphor means representing a metaphor with your body. For example, the metaphor thinking outside the box, which means thinking in an open and unconstrained way, was embodied by having people think about solving problems either as they walked around the perimeter of a square for two minutes (thinking inside the box) or by walking freely, unconstrained, about the room for the same period of time (thinking outside the box). Remarkably, they found that thinking was affected by the way people walked. Those who walked freely outside the box scored higher in two different tests of creativity, which measured originality of thought, than those who were constrained and had to walk around the perimeter of a box. On the basis of five experiments, Leung et al. conclude that embodying creative metaphors enhances creativity. They account for this by suggesting that normal creative thinking becomes rigid and fixed and that moving our body freely and without constraints "may inhibit unconscious mental barriers that restrict creative cognition and thereby boost performance on both convergent and divergent thinking tasks" (p. 507). So, the idea here is that moving freely can get rid of "mental barriers" that stop us being creative. It follows from this argument that if someone is already fulfilling their maximum creative potential then there is nothing more they can do, but if they have potential to spare then moving freely might release them from unwanted cognitive constraints.

When Leung et al. talk about convergent and divergent thinking they are referring to two different types of problem solving tasks. To illustrate each type of problem solving task lets work through some examples.

Convergent thinking

Let's start with a simple bit of arithmetic. What's five times three? Easy, it's 15 and you probably knew the answer without having to think about it. What's 15 times three? Still easy, it's 45. Again, you might have thought for a moment but then found the answer without having to do too much work. Now, what's 43 times nine? This one generally takes people a bit longer to work out. Once you've worked out the answer (in your head if you can, without writing it down or using a calculator) think about how you worked it out.

Some people work it out using the following steps:

$$\text{Step 1: } 43 \times 10 = 430$$
$$\text{Step 2: } 430 - 43 = 387$$

Other people use a different set of steps:

Step 1: 3 x 9 = 27
Step 2: 40 x 9 = 360
Step 3: 27+360 = 387

Whichever way you do it, you are finding the answer by going through lots of discreet steps. If you do this inside your head then you are finding the answer by using convergent thinking. In everyday problem-solving tasks we use convergent thinking when we are trying to find the one, single, "right" answer to a particular problem. A simple example of when we use convergent thinking in everyday life is when we are following a set of steps for a recipe that we've remembered.

Divergent thinking
Here's a different sort of problem with lots of right answers rather than just one. I'm going to name a common object and I want you to think of as many uses for it as you can. An alternative use is a use for the object for which it was never designed. Here's an example. An alternative use for a sheet of paper might be to fold it up into a thin point and then use it to pick your nose. The sheet of paper wasn't designed as a nose-picker but you can certainly use it to pick your nose, if you're that way inclined.

The task is to think of as many different uses as you can, as quickly as you can. Can you think of at least seven uses for a regular household brick? Some people find this task really hard. They can think of three or four alternative uses but then start to struggle to think of more. People tend to get stuck, thinking about a particular alternative use and they can't move on. Why don't you write down the alternative uses for the brick that you thought of?

This type of problem solving is called divergent thinking because you have to think outside the box, you have to diverge from what is known or expected. Of the hundreds of alternative uses for a brick I have heard, the two that stick in my mind are, to rub a wart off your skin and to turn it upside down and use it as an ashtray.

We use divergent thinking when we are trying to solve problems that have more than one correct answer. For example, the problem of what meals to cook for dinner on different nights of the week, or drawing up a short-list of where to go on holiday next year. Planning your weekly food menus and your holidays require divergent thinking because there are lots of different meals that you could prepare and there are lots of places that you could visit on your holidays. However, we tend not to use very much divergent thinking when we are making such plans. Most of us tend to eat the same food week in week out and I am sure that most of us have visited the same place on holiday more than once, even though there are thousands of alternative places that we could, yet haven't visited. For that reason, it's probably the case that whereas deciding what to have for dinner next Thursday should involve a degree of divergent thinking the reality is that we are more likely to decide on our menus based on convergent thinking styles (it's Thursday, therefore it must be lasagna). And herein rests the problem. We find it hard to do something different, to be creative.

Dance and divergent thinking

Maxine Campion and Liat Levita carried out a neat study into the relationship between dance, exercise, mood and creativity (Campion & Levita, 2014). They asked participants to carry out tests of mood and creativity before and after they spent just five minutes either listening to music, dancing, cycling or sitting quietly. While both the dancing and the cycling were physiologically arousing, and led to similar changes in heart rate, there were distinct differences in people's mood and creativity, which changed depending on which activity they did. For those in the dancing group there were significant increases in positive affect (such that participants felt more happy) and significant decreases in negative affect (such that participants felt less unhappy). People in the dancing group also showed significantly reduced feelings of fatigue. For those in the cycling group there were no significant changes in either positive or negative affect (mood) and there was no reduction in feelings of fatigue.

With regards to creativity Campion and Levita (2014) found that for those in the dancing condition there was a significant associated between mood and creativity, such that increases in mood were associated with increases in verbal creativity. It seems to be the case, based on the findings of this study, that while five minutes of dancing and cycling can lead to similar physiological changes, only dancing leads to improvements in mood and associated positive changes in verbal creativity and divergent thinking.

Why would this be the case? One of the explanatory frameworks examined by Campion and Levita (2014) to account for these findings is Fredrickson's (2004) Broaden-and-Build Hypothesis, which states that improvements in mood lead to broadened cognitive functioning. In other words, increases in creativity will arise from increased mood; such that the happier we are the more creative we'll be. In another possible explanation they point out that moving freely to music might enhance both mood and creativity. While the same pop song, Do Your Thing by Basement Jaxx, was played for both the dancing and the cycling conditions those people in the dance condition were asked to dance along to the rhythm of the music whereas those in the cycling group were asked to just cycle and not coordinate their cycle movements with the rhythm of the music. Perhaps it is the case that entraining physical movements to rhythmic beats (moving to the rhythm of a beat) increases both mood and creativity and reduces fatigue.

There is clearly a link between moving and thinking. When we move our body it changes our thought processes, either by speeding up the way we think, or by making us think more creatively. Of course, not all types of dancing (or moving) are the same. One form of dancing (like the freestyle form of dancing people do in nightclubs) is very different from another form of dancing (like the country dancing done by Morris Dancers). This led me to question whether different types of dancing had different effects on the way people think. For example, I wanted to know whether there was one type of dancing that made us more artistic and creative and another form of dancing that made us more careful and conservative.

Carine Lewis, one of my former graduate students, and I set up an experiment in our lab (see Lewis, 2012). We invited people to take part in a study that was to do with thinking and problem solving. When people arrived at the lab they were given a battery of divergent and convergent thinking tests. Some of the tests were visual, some were number-based and some were language-based.

Mental rotation task

For one of the visual tests we used the Mental Rotation Task (see Shepard& Metzler, 1971). In this task you are shown drawings of two 3D objects, but from different angles, and you have to decide if the two images are the same or different.

To solve this puzzle you have to mentally rotate one of the images around in your mind until it matches the other image. If the images match then they are the same. This is a convergent thinking task because there is only one correct answer (the images are either the same or they are different).

Numbers

The second example comes from a mathematics based divergent thinking task. In this task you are given a set of numbers and the operators: plus, minus, multiply and divide ("+", "-", "x" and "/"). You are then given a target number. Your task is to combine the numbers and operators in such a way that you are able to end up with the target number. In this version of the task you have to find as many different ways of reaching the target as possible. Therefore, once you've found one answer you have to find another. Here's the example:

Numbers	Target
2, 6, 8, 12, 24	74

How many ways are there to make 74 from these numbers? Here are some solutions:

Solution 1: 8 x (12 – 2) - 6
Solution 2: 2 + (6 x 12)
Solution 3: (12 + 24) x 2 + (8 - 6)
Solution 4: (6 x 8) + (24 +2)

Structures and improvised dance

We then created two dance-instruction videos. Both of the videos were of the "follow-me" format. I demonstrated what the participants needed to do and the participants generally copied, or did what I told them to do. In the first video, I led the participants in a structured dance routine. I moved my arms, legs and torso, stretched high into the air and low to the ground and used a range of movements that were simple and which were coordinated to the regular beat of the music. In the second video, I led people in an improvised dance session.

Improvisation involves moving spontaneously in an unplanned way. This is hard. We are all so caught up in our own set patterns of behaving, moving and thinking that it is hard for us to think of new ways to move. In the improvisation video, I encouraged people to focus on one part of their body, for example their arms. I asked people to copy my improvised arm movements, so they could get the hang of things, and then I stopped moving but the people watching the video had to carry on moving and to keep on changing their movements until I told them to move a different body part or move in a different style. The video encouraged people to dance as robots, as leaves floating on the wind, and as heavy rockers. Half of our experimental participants danced to the structured dance video and the other half danced to the improvised dance video and then they completed a second version of the battery of cognitive and problem-solving tests.

When we analysed the results of our study we found two really interesting things about how dancing in different ways has an effect on the way people think. First, we found that the people who had danced to the structured dance video became faster at solving convergent problem-solving puzzles after they had danced. A simple example of one of the convergent problem-solving tasks we gave people was the lexical decision task. In this task people are given a string of letters, such as "FAMERY", and they have to decide whether the string of letters is a real-word or a non-word. They do this by pressing one of two buttons on a computer keyboard and they are encouraged to do it as quickly as they can. People can normally do this in a few hundred milliseconds and the task measures how quickly people are able to mentally process visually presented words. What we found was that dancing for 20 minutes, in a structured way, made people faster at mentally processing information. What was even more remarkable was that even though people got faster at processing this information and finding the solution, they did so without any loss of accuracy. In other words, dancing in a structured way helped people in this study to find the single, correct answer to a problem faster.

When people danced along to the improvisation video there was no increase in their speed of thinking. It seems to be the case that structured dance makes people think faster but improvised dancing does not. However, improvised dancing did have an effect on a different aspect of problem solving. What we found was that after 20 minutes of improvised dancing people became more creative in the answers they gave to divergent thinking tasks. For example, before dancing people could generate about 4 or 5 alternative uses for a common object such as a brick or a newspaper, but after dancing in an improvised way they could generate 7 or 8 alternative uses. This is nearly a 100 percent increase in output. While improvised dance in our study helped people to think more creatively, structured dance did not.

Thinking in set patterns

We tend to think using set patterns of thinking. This is important for us for several reasons. It enables us to predict what we are about to see and hear, it gives us a guide as to how to behave when we go somewhere new, and it helps us to resolve ambiguity as we process information from our surroundings. Set patterns of thinking are learnt with experience and we build them up as we go through life. They are generally good things to use because they mean we can go through life without having to think too much. For example, let's imagine you are listening to someone you know speak on a poor telephone line. Even if the signal is fairly degraded and you can only hear parts of what they are saying you may be able to piece together the meaning of what they are saying because you are able to use the context of the conversation and knowledge of the person who is speaking to "fill in the gaps". In this situation, you are using set patterns of thinking to disambiguate, or fill in missing information. However, not having to think too much has its downside too. It means that we might think we know what is going on when really, we have got completely the wrong end of the stick. It means that we are not very good at dealing with new things, things that don't fit into one of our set patterns. When we rely too heavily on set patterns of thinking we find it difficult to be creative and process new information. Set patterns of thinking allow us to think and act automatically. In fact, set patterns of thinking allow us not to think. The reason improvised dance helps us to be more creative is because it is helping us to break away from set patterns of movement, which in turn helps us to break away from set patterns of thinking. For further work on improvisation and thinking see Lewis & Lovatt (2013). There is, clearly, a very strong link between our movements and our thinking

Dance is about movement and it's also about moments of stillness. It's about projecting character through body posture and articulating feelings through physical expression.

Strike a Pose

In 1990 Madonna released a song called Vogue whose opening line is "Strike a pose". The basic premise of the song is that if you want to escape the pain of life and be something else, something better, then you should lose yourself on the dance floor. The idea being that striking a pose can help you unlock your imagination, where you'll find inner beauty and inspiration. Of course, this is just a pop song. Vogue was a huge international hit and the music video that went with it showcased a style of nightclub dancing called "vogueing".

Several scientific studies have been published which seem to show that striking a pose (or standing in a particular way) can change the way people think and feel (Carney, Cuddy & Yap, 2010; Huang, Galinsky, Gruenfeld & Guillory, 2011; Riskin & Gotay, 1982; Tiedens & Fragale, 2003).

Carney, Cuddy and Yap (2010), researchers at Columbia and Harvard Universities, carried out one of the studies and what they found blurs the distinction between scientific discovery and artistic expression, because it confirms that what artists know intuitively scientists have to "discover" in a lab. Madonna's lyrics suggest that striking a pose might change the way we think about and see ourselves and this scientific study has found, empirically, that not only does striking a pose change the way people think it also changes people on a physiological level too.

Forty-two people were divided into two groups and Carney et al. made them stand or sit in a number of poses. The people in the first group were placed in "high power poses". A high-power pose is a way of standing or sitting that looks confident, relaxed and self-assured. Imagine leaning back in your chair, with your arms behind your head and your feet up on the desk in front of you. Or leaning over a desk with your arms spread wide. These are high power poses. The people in the second group were placed in "low power poses", which are poses that look closed and constricted. Imagine sitting on a chair with both feet flat on the floor with your hands crossed over your lap, or standing with your legs and arms crossed. These are thought to be low power poses. So, half the participants were placed in high power poses for a total of two minutes and the other half were placed in low power poses for a total of two minutes.

Carney et al. then measured three things. First, they asked the participants how "powerful" and "in charge" they felt and those people who had been standing or sitting in high power poses said that they felt more powerful and more in charge than those who'd been standing or sitting in low power poses. Second, they set the participants a "risk-taking" gambling task. Imagine this. You're given ten pounds and a dice. You can either keep the money (the safe bet) or you can roll the dice (the high-risk bet). If you roll the dice and you get either a 1, 3 or 5 then you lose all the money. However, if you roll a 2, 4 or 6 then you double your money from ten pounds to twenty pounds. What would you do? The scientists found that 86 percent of those people who had been placed in the high-power poses took the high-risk bet, whereas only 60 percent of the people placed in the low-power pose took the high-risk bet. It seems to be the case that just standing or sitting in a particular pose for a couple of minutes is enough to change how powerful people feel and it also influences their risk-taking behaviour. This is interesting, but what I find really amazing is the researcher's third finding.

Before the participants took up their poses they each had to provide a saliva sample. These saliva samples were used to measure concentrations of the stress hormone cortisol and the sex hormone testosterone in each of the participants. It is thought that high testosterone is a marker of high dominance in people and animals and that a high level of cortisol is a marker of a low-power individual. Once the participants had finished posing they provided a second saliva sample. What I find incredible, almost unbelievable, is that for people in the high-power posing group their testosterone levels went up and their cortisol levels went down. Whereas the complete opposite happened in the people in the low-power posing group. Their testosterone level went down and their cortisol level went up.

So, stationary postures not only affect how we think about ourselves but they also influence our hormonal state. Standing or sitting in a particular way for just two minutes has a profound effect on us.

Summary

There appears to be a connection between our body and the way we think and solve problems. Moving our body in different ways seems to change our ability to solve different types of problems and the simple act of standing in different positions influences not only how we feel and behave, it also influences our hormonal state. I do wish my teachers had let me fidget more in school, perhaps I might have left with better grades.

CHAPTER 3

Dance, Mood and Depression

I have asked over a thousand people the simple question "Why do you dance?" I've asked men and women from right across the life span, from children as young as 6 to adults in their late 70s. A large number of the adults told me they dance to make themselves feel happy and overcome feelings of depression. One woman told me how dancing keeps her from feeling depressed and makes her forget about the stresses of everyday life:

> *"Dancing makes me happy and makes me forget about daily stresses and things that bother me. I've danced my entire life and I've tried all types of dancing. I will never stop dancing. It keeps me from feeling down and depressed and it gives me energy..."*
> (Female, 20)

Another woman, in her early 60s, told me that dancing transforms her from being grumpy and miserable to being happy and positive. Even though she isn't as physically mobile as she once was she told me that she can't stop dancing:

> *"I dance because I cant stop, even if I can't do a proper class anymore (mashed up knees) dancing changes me from miserable,*

grumpy old lady to positive and happy sexy woman. I dance when I am down and I can change my mood completely. I love to go out and dance and dance all night." (Female, 62)

One of the features of depression concerns the way people think about themselves. Depressed people can have negative thoughts and these thoughts can feel like they're filling up a person's head. This can be overwhelming as it becomes increasingly difficult to switch these thoughts off, so that you have sufficient head-space to think about other things. People have told me that dancing helps them to switch off their mind. Having a time where you can be free of overwhelming or negative thoughts lifts the weight of anxiety from a person's shoulders and gives them a feeling of increased vigour and happiness. Here is what one man in his late twenties told me about how dancing helps him move his focus from his mind to his body and how this makes him feel excessively happy:

"Dancing makes me inordinately happy. Every time I go dancing I wish I would dance more often. I spend a lot of time thinking and when I'm dancing I think I'm truly in my body rather than in my head. It's brilliant." (Male, 27)

People who know the secret of dancing, that dancing has the power to unlock positive feelings and make them feel happy, know that it is a tool that can be used throughout their life to banish depressive feelings:

"The reason that I like to dance is still the same as when I was younger: these are the only moments I feel happy and I can't stop smiling." (Male, 54)

One woman in her late 50's has described to me how dancing during troubled times was her salvation. Having not danced for several years she described how her body felt toxic and over-weight and how dancing helped restore feelings of happiness and confidence.

> *"I have always loved to dance and move my body. In troubled times dance has always been my 'salvation'. There was a 6-year period where I didn't dance because my husband didn't. I felt toxic and gained 50 pounds. Now I dance almost every night. I'm slim, happy, confident, and wouldn't be without the release of dance."*
> (Female, 59)

People tell me over and over again how dancing makes them happy. Dance is a wonderful anti-depressant because it naturally addresses most of the major sets of symptoms associated with depression, which I describe below.

Depression

Depression can be defined as a state of low mood, or of being low-spirited. People who experience depression often show an aversion to activity, because every-day activities feel harder and less worthwhile. The symptoms of depression can be understood in terms of four categories.

Mood

A person with depression might have low mood, or the blues, much of the time and they might find no pleasure in things. They might feel sad and hopeless and not respond in a positive way to those things and people that used to make them happy. Someone with depression might say, for example, that there is no happiness, just the absence of pain and discomfort.

Behaviour

Low mood and a failure to find pleasure in things can also lead to a change in a person's behaviour. A person with depression might stop doing activities that they once enjoyed and they might start to avoid social events. This can lead to inactivity and social isolation.

Thinking

There is a connection between the way people feel and the way they think. People with depression might lose confidence in themselves, and this can lead to negative changes in their self-esteem (how much they value themselves). People might also find it difficult to concentrate on things and they might start to notice changes in their memory. For example, in some cases memories can be mood dependent. This means that when we are in a low mood we are more likely to remember and bring to mind negative thoughts and when we are in a positive mood we are more likely to remember and bring to mind positive thoughts.

Physical

Depression doesn't just effect the way people think and feel it also effects their physical state too. For example, people with depression might have a lack of energy, feel tired and lethargic and move around slowly. They might find it hard to pull themselves out of a chair and do something. Despite this they might have real difficulty resting and ultimately sleeping.

The severity of a person's depression is classified on a scale from sub threshold to severe, based on how many symptoms they have and their degree of functional impairment (functional impairment is based on a person's need for assistance to carry out specific activities – activities they could do on their own before the onset of depression). Somebody with the mildest form of depressive symptoms (sub threshold) may have very few symptoms and no functional impairment. Mild depression is characterised by having five or more symptoms of depression and minor functional impairment. Moderate depression is between mild and severe and severe depression is characterised by having most of the symptoms of depression and includes a high degree of functional impairment.

Treatments for depression vary, depending on the severity of the depression and on people's individual needs and preferences. Treatments might include different forms of CBT (Cognitive Behavioural Therapy), drugs, ECT (Electroconvulsive therapy), one of the many talking therapies and exercise.

Exercise and Dance

Exercise, that is done regularly, can have a positive effect on low mood and low energy levels, such that it can lead to improvements in mood and help people to feel an increase in vigour. Regular exercise can also help people to get off to sleep at night. The value of exercise for helping people to overcome feelings of depression is recognised by many family doctors, some of whom give out prescriptions so that people can attend exercise sessions.

Dance is a great form of exercise because it works on so many different levels and taps in to the four classes of symptoms associated with depression. Dance has been shown, in scientific studies, to improve a person's mood, it is a behaviour that everyone can do, regardless of their age, and requires no specialist equipment, clothing or facilities. Dancing, and learning a dance sequence, taps into a person's thinking skills, such that it encourages them to concentrate, learn and remember new things and it can be done either alone or as part of a small or large group. There is evidence that dancing has a positive impact on the symptoms of depression.

Studies on Dance and Mood

Dance has been shown to lift the mood of people with both mild and severe depression. In one study, carried out in Germany by Koch, Morlinghaus & Fuchs (2007), examining the effects of dance on people who had been admitted to a psychiatric hospital, all of whom had been diagnosed with depression, it was found that just one 30-minute session of dance was enough to reduce the symptoms of depression and to increase feelings of vitality (a feeling of being full of energy, strong and vital). The dance that was used to lead to these changes in mood was the Hava Nagila dance. Danced in a circle, it is lively, bouncy and upbeat. Hava Nagila is Hebrew, meaning "let us rejoice". The music that accompanies the dance is very popular and uplifting and the researchers wanted to know whether it was the music, on its own, that was causing the reduction in depressive symptoms. So, they ran another condition where a second group of patients with depression just sat and listened to the music. In this music-only group, the patients actually became slightly more depressed! So, it seems to be the case that dancing in an upbeat, lively and bouncy way, independently of music, can have a positive effect on feelings of depression and vitality. It seems remarkable, also, that just one

30-minute session of dance is sufficient to lead to observable reductions in depression.

In another study, this one carried out in Korea, Jeong, Hong, Lee, Park, Kim & Suh (2005) wanted to know whether a longer-term programme of dance would lead to improvements in mood in a group of 16-year-old girls who had mild depression. The girls in this study were regular school girls and had not been admitted to hospital. The girls were divided into two groups. One group took part in 3 dance sessions a week for 12 weeks and the other group did nothing. The dance sessions were focused on aspects of body awareness, movement, and expressing feelings and images, and were based on dance-movement therapy techniques. The researchers found that dancing 3 times a week for 12 weeks led to a reduction in feelings of depression, anxiety and hostility for the girls who danced but there was no reduction in these aspects of mood for the girls who did not take part in the dance sessions. The explanation given for this reduction in negative psychological symptoms is that the 12 weeks of dancing may have its effect through physiological changes, such as the girls becoming more physically relaxed, or through changes in the concentration of stress hormones circulating around their bodies.

What is interesting about both of these studies is that different types of dancing, something lively and energetic in the first study and something reflective and expressive in the second, both have a positive impact on the mood of people with either severe or mild depression.

Of course, people who are depressed have a lower state of mood than they would have if they were not depressed. In other words, during a depressive episode, mood levels for an individual will go down. When people who are depressed dance it seems that their mood levels go up again, perhaps back towards their pre-depression levels. This leads to an interesting couple of questions about dancing and mood. Is it the case that dancing only improves the mood of people who are depressed? And, is it possible for people who are not depressed to have their mood improved by dancing? A couple of studies by Andrew Lane, from the University of Wolverhampton, help us to shed some light on this.

Andrew Lane's first study (Lane & Lovejoy, 2001) suggests that dancing can help to improve the mood of people who are both depressed and not depressed, but that dancing has the biggest effect on people's mood when they are depressed. Lane & Lovejoy (2001) gave 80 people a mood questionnaire which measured moods such as tension, anger, fatigue, depression, vigour and confusion. From the results of this questionnaire people could be grouped according to how depressed they were. Some people were classified as depressed and some people as not. Once the questionnaire was completed, everyone took part in a 60-minute aerobic dance session. After the dance-session the questionnaire was completed again. Using this method, Lane & Lovejoy (2001) were able to address two questions: whether the dance session led to a change in mood and whether this change was greater in people who had been classified as depressed than in those people who had been classified as not depressed at the start of the class. The results showed that, following the dance class, there was a general reduction in feelings of anger, confusion, fatigue and tension across all participants. When the results of the depressed group were compared with the results of the non-depressed group, it was found that the reduction in feelings of anger, confusion, fatigue and tension were greater in the depressed group.

Mood Studies with Dancers in Vocational Training

All the scientific studies that I've mentioned so far have looked at the effect of dancing on the mood of people who are not professional dancers. This leaves open the question of whether you still get a change in mood even when you perform or practice dance lots of times. It's possible that when people dance a lot they experience less mood changes than people who dance less frequently. Andrew Lane's second study helps us to clear this up. With colleagues from the Laban Dance Centre in London and the British Olympic Medical Centre, Lane, Hewston, Redding and Whyte (2003) measured the mood of dancers in full-time training before and after two different dance classes. The two dance classes were for different types of contemporary dance. One class was based on the technique of Martha Graham and the movements performed were based on contraction and release; they were often angular, static and close to the ground (just as an aside, when I was training as a dancer I really didn't enjoy dancing this style). The other class was based on the technique of José Limón and are based on the principles of tension-free movements with a sense of flow (from my own experience, the Limón technique feels nice, flowing and natural to dance, whereas the Graham technique felt awkward, clumsy and unnatural). The results of this study were different from the results of the other studies on mood and dance. In this study only one aspect of mood, vigor, increased as a consequence of dancing and even then, this change was only observed following the Limón class. What this finding suggests is that for full-time dancers some dance styles lead to a change in some moods. It's not very decisive, but that's the way science turns out sometimes, and we are left with more questions than answers. In this case we are left asking which of the hundreds of dance forms that people regularly engage with lead to changes in which emotions and feelings?

Parkinson's disease, Dance and Changes in Mood

As the head of the Dance Psychology Lab I'm always on the look-out for scientific research related to dance. I came across some research by a group of physiotherapists in the USA (Hackney, Kantorovich, Levin & Earhart, 2007) which showed that when people with Parkinson's disease took part in tango dance classes they showed a significant improvement in balance. This is important because poor balance is one of the negative symptoms of Parkinson's disease. What was even more striking for me was that the researchers had either given people with Parkinson's disease 20 sessions of tango classes or 20 sessions of exercise classes and they found an improvement in balance only for those in the tango group. This suggested to me that there is something special about dance that provides a benefit over and above that derived from exercise. Over the next few years several other researchers reported similar findings and I wanted to see whether dance could also have a moderating effect on some of the psychological symptoms associated with Parkinson's disease, such as mood and problem solving.

Working with a team of researchers at the University of Hertfordshire, who each brought something special to the project (Dr Lucy Annett – Neuroscientist, Sally Davenport – Physiotherapist, Dr Carine Lewis – Cognitive Psychologist, Dr Amelia Hall – Psychologist, and myself a Dance Psychologist) we were interested in the effect of recreational dance on the mood of elderly people, and more specifically on older people with

Parkinson's disease (Lewis, Annett, Davenport, Hall, & Lovatt 2016). We recruited a group of 37 people aged between 50 and 80. Twenty-two of the participants had been diagnosed with mild to moderate Parkinson's disease and the other fifteen acted as age-matched controls. Our participants took part in weekly dance sessions for ten weeks. The dance sessions were designed to be upbeat and characterful and the dances could be done either sitting down or standing up. In total, the participants learnt five dances based on Bollywood, Tango, Cheerleading, Old Time Music Hall and a medley of party dances, which included the Charleston and Saturday Night Fever. The style of dancing was changed every two weeks.

Our participants therefore had to learn a new dance routine and remember it every two weeks (this therefore required the cognitive processes of learning and memory). They had to execute the dance routine (requiring the cognitive processes of spatial awareness and the exertion of physical activity) and they had to move in time to music (requiring the process of sensori-motor coupling, whereby they had to synchronise their movements to musical rhythms). We measured the mood of our participants using two instruments: The Profile of Mood States (POMS) and the Brunel University Mood Scale (BRUMS). We were interested in both short-term and longer-term changes in mood. Therefore, we measured short-cycle changes in mood by taking measures of mood immediately before and immediately after a dance class and we measured long-cycle changes in mood by taking measures of mood a week before the first dance class and a week after the final dance class. We found that total mood disturbance changed significantly (in a positive direction) over time, both in the short-cycle and the long-cycle. In general, participants felt less tense, and for those people who had higher levels of depression they reported less fatigue following the dance classes.

We concluded that taking part in weekly dance classes can improve mood in the elderly, with or without Parkinson's disease. However, we are unclear about the variables and/or mechanisms that might influence the magnitude of the effect of dancing on mood. There are many different aspects to dance. For example, our dance sessions were very social, and so part of the improvement in mood might be attributable to the social aspect of dance. Our dance sessions were also safe, caring and unthreatening and it may be this environment that partially led to improvements in mood. This is known as the "healing balm" effect. Finally, all of our dance sessions were accompanied by lively popular music, which can make you want to tap your feet, and so this might also have contributed to improvements in mood. In future studies, we hope to tease these aspects apart.

Dance seems to have a positive effect on mental health and studies, including our own, show the greatest benefit for people with depressed mood. Studies have shown that dance can help to increase feelings of vigour and decrease feelings of fatigue. Therefore, if you're feeling tired or you've got low energy levels or you're feeling a little depressed, then dancing might help you feel more awake, full of energy and put you in a more positive state of mood.

Summary

As a society, we spend billions of pounds every year on health care. If we can establish that dance has a predictable and reliable effect in terms of helping to reduce some of the symptoms of ill-health then prescribing dance could become part of a treatment plan for some people. Dance is inexpensive, easy to do and entirely natural. Our study has shown that dancing can improve the mood of older people. Previous studies have shown that dance can be used as part of a weight management programme, been effective at helping people with arthritis and is an effective tool for delivering the exercise necessary following cardio vascular disease. Having an understanding of how moving our bodies changes our physical and psychological health and well-being would be very useful to society.

CHAPTER 4

Trance and the Dancing Plagues

When people go into a trance, or a trance-like state, they experience a change in mind-set, which goes far beyond a simple change of mood. A trance resembles a hypnotic state, or perhaps a sense of feeling very sleepy. When you are in a trance you are less aware of what's going on around you and some people say that when they're in a trance they lose consciousness. It's hard to pin down exactly what is happening when people experience a trance-like state but it's a bit like that time first thing in the morning when you're half way between being asleep and awake, when your arms and legs don't feel like they belong to you, and you're not sure if you're still dreaming or awake and you can't work out where you are. Being in a trance is like being in a semi-conscious daydream.

When people are in a trance-like state it is thought they can experience a feeling of dizziness or of vertigo and this can be accompanied by a sense of disappearance or loss of self. This means that people stop thinking about themselves as others see them and instead they start to become introspective and, if they enter a trance-like state with other people, they can start to experience a perception of collective consciousness. In a trance-like state a person's focus of attention can become narrower and this might be accompanied by changes in bodily perception. So, in one sense, a trance-like state is an altered state of consciousness, a fundamental change in conscious mind-set. A lot has been written about trance but the question for us is whether the act of dancing can lead to a state of trance. Can moving our body lead to a fundamental change in conscious mind-set? If so, what are the necessary conditions?

Historical Accounts of Trance Dancing

There are several historical accounts which link dancing with trance-like states. Although these accounts are interesting they are not based on scientific research, and because of that they do not provide the detail that we require to answer specific questions about the link between dance and trance. Nevertheless, let's have a look at them.

Frau Troffea and the dancing plague

In 1518 something strange seems to have happened in Strasbourg. A woman called Frau Troffea started to dance alone by a river; she carried on dancing, non-stop, for several days. Eventually hundreds of other people started to dance with her and they all danced together until their feet began to bleed. They carried on dancing until dozens of the dancers died. This phenomenon is called a dancing plague: the dancing seems to be contagious and the people who catch the dancing can eventually die from it. It seems incredible but, apparently, this wasn't an isolated incident. Many other dancing plagues have been reported (from as early as 1017), which seized and killed men, women and children. The people who caught the dancing plague were thought to be in a state of trance because reports suggest that the dancers appeared to lose self-control. The dancing was relentless, lasting for several days, and it was sometimes frenzied. It is not clear how any dancing plague starts. One suggestion is that the dancers were all struck down with a bout of food poisoning (a particular kind known as ergotism caused by mould on grains) which caused them to hallucinate and dance in a wild manner. Another suggestion is that the dancing was a response to repressive social systems and triggered by social catastrophe and hardship. We do not know whether the dancing caused the trance-like state or whether the dancing was a side effect of whatever caused the trance. Nevertheless, it seems that during the Renaissance people danced in a trance-like state. John Waller has written an excellent book (Waller, 2009) on the dancing plagues, called *The Dancing Plague: The Strange, True Story of an Extraordinary Illness*

Typical, of course, of this kind of historical report of dancing is that we do not have a reliable record of exactly what was meant by "dancing" during the dancing plagues. Was the dancing choreographed, with everyone doing the same moves (Thriller-style) or was the dancing random, such that it was unpredictable and uncontrolled? Perhaps the dancing was at times frantic and energetic and at other times gentle and flowing? Did everyone dance in a particular group dance in a similar way (as we see in nightclubs today) or was the dancing an expression of the kind of individuality that you might expect to see if people were dancing completely on their own? Perhaps the dancing was simply a repetitive rocking motion, as is seen in modern dance marathons or perhaps a type of bouncing that is seen in some traditional African dances.

The Psychology of the Dancing Curse

One account of the dancing plagues was that people were afflicted by dancing curses, that were sent down from heaven or up from hell. From the 15th to the early 18th century there was widespread belief, in northern Europe, in the possibility of divine curse, bewitchment and possession. Within this context it might have been believed that someone dancing in a trance-like state could have been possessed, and as such might provide evidence for the presence of a satanic spirit nearby. Others may have joined the dancing because they too may have believed that they were possessed by the satanic spirit. As more people joined in the dance it would have provided greater evidence of the presence of a dancing curse. This type of logic is known as a confirmation bias, where people look for evidence which supports their belief. In this case people believe in possession and in the power of the spirits to send down dancing curses. Therefore, when they see someone dancing in a trance-like state this "confirms" their belief which makes them susceptible to the suggestion that they too may become possessed. Now, if there really is something about the pure act of dancing that is contagious (for example, other people's movements, the rhythm, a shared compulsion to dance) then coupled with such a belief system it is easy to see how dancing plagues spread during the renaissance period.

Modern Day Dance Trance States - DJ-induced trance

The rave culture of the 1980's and 1990's may seem a long way from the dancing plagues and curses of the 16th century but they share many elements of dance and trance. In the underground rave clubs of the 1980's and 1990's people would dance in a group, altogether, for hours at a time. Although people at raves didn't dance for as many days as those who reportedly danced in the dancing plagues, the number of people who danced together at raves was much higher.

In one report by Gore (1997) it is suggested that raves used to attract between 8,000 and 10,000 people and other rave-based events, like the Berlin Love Parades, would sometimes attract over a million people. The way in which people danced at raves was influenced by electronic dance music; such as techno, house and acid, and the speed of the rhythm at which it was played. Typically, rave music was played at an average speed of about 120 - 130 beats per minute (bpm). This is interesting because it's about the same rate at which the heart of a 20-year-old beats when they're engaged in moderately intensive exercise. At this bpm healthy young dancers can keep dancing for long periods of time.

Rave DJ's would control huge crowds of dancers by alternating the rhythm and bpm of the music they played throughout the evening. Good DJ's were able to bring a crowd close to a frenzied climax by raising the speed of the music to between 160 and 220 bpm and then bring the dancers down again to a manageable, maintainable, 130 bpm. The effect of raising the bpm to 160 – 220 is that, to keep up with the beat, dancers must be working at close to, or even exceeding, their maximum heart rate (which for 20 year olds is about 200 bpm). At this bpm the dance moves of ravers become more like spontaneous, automatic, muscular twitches than anything planned or choreographed and it is at this stage that rave dancers are thought to enter into a trance-like state where the body moves as a direct response to the music, such that the sensorimotor coupling of music and movement is more like an automatic startle–reflex than a choreographed or planned-in-advance set of movements. A state without time for conscious planning. At this stage, the dancer is completely under the control of the music and, ultimately, the person controlling the music, the DJ. Once in a DJ-induced trance there is little definition in time, just the constant beat of the music.

Summary

Of course, the question still remains. Is it the dancing that leads to the trance-like state or is it something else, like the music? Just because dance and trance are observed together we cannot infer that it is the dancing that causes the trance. All we can say at this stage is that dancing is associated with the altered states of consciousness (mind) that people go through when they are in a trance-like state. In the examples I've given so far, it seems to be the case that people start to dance first, both in the case of the dancing plagues and the rave scene, and then they enter into a trance-like state.

CHAPTER 5

Dance Confidence across the Lifespan

Imagine you're at a party, nightclub or wedding where you're dancing along with other people. Imagine you're dancing to your favourite piece of music. You're in your groove and feeling great. As you look around the room you notice that everyone else who's dancing is the same age and gender as you. How are they dancing? Now, answer the following question. Compared to other people of your own age and gender how good a dancer do you think you are? I don't mean how accomplished you are in terms of technique or achievements in a particular style of dance (such as ballroom or lindy hop), I mean how confident do you feel about just letting yourself go and moving to the music? Give yourself a score between 1 and 7, where 1 is terrible and 7 is fantastic.

The way you answered this question will have been influenced by many factors. However, there are two things that will have had a major effect on your answer. These are your age and your sex. In a survey of nearly 14,000 people I have found (Lovatt, 2011) that a person's dance confidence (which is what I am measuring with this question and which is part of what makes up your self-esteem) is influenced by their age and sex. Moreover, changes in dance-confidence-related self-esteem happen at interesting points across the lifespan and these are different for men and women.

The major differences in dance-confidence self-esteem between men and women across the lifespan can be seen in the following graph.

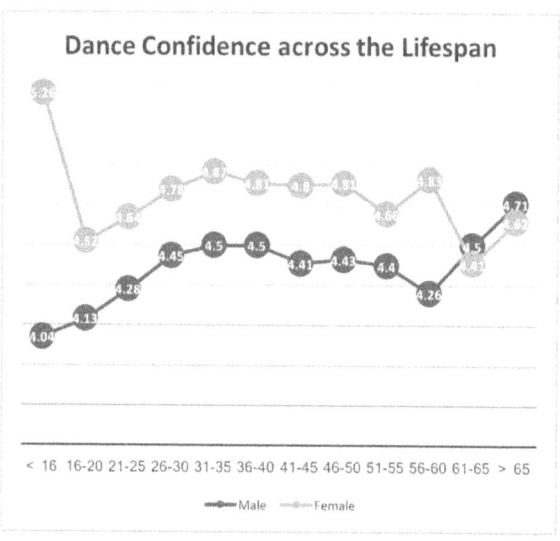

Dance Confidence across the Lifespan

Self-esteem and Dance

Before I tell you about what I found out, I want to tell you about two models of self-esteem, which might help to explain why my measure of dance-confidence self-esteem changes with age and sex.

The first model is the *Competencies Model* (James, 1890). According to this model, someone's self-esteem in a particular area of their life will be influenced by how competent (or good) they feel they are at doing something. So, with regards to dance-confidence self-esteem, if someone feels that they are a competent dancer then they will have higher dance-confidence self-esteem than someone who feels that they are a less competent dancer.

The second model is the *Reflected Appraisal Model* (Baumeister & Leary, 1995). According to this model, someone's self-esteem in a particular area of their life will be influenced by how they think other people view them. So, with regards to dance-confidence self-esteem, if someone feels that other people think they are good at social and recreational dancing then they will have higher dance-confidence self-esteem than someone who believes other people don't think they are very good at social and recreational dancing.

My Study of 14,000 People

Some people think that psychology is just the science of the obvious and that my finding of sex differences in dance-confidence self-esteem, specifically that women have higher dance-confidence self-esteem than men, is just that – obvious. You might be thinking that I didn't need to ask 14,000 people to find this out. We can look at virtually any group of people and see that women are generally more confident about dancing than men. However, by asking that number of people I was able to see some really interesting patterns that are less obvious and more surprising.

I found (Lovatt, 2011) that women under the age of 60 have higher levels of dance-confidence self-esteem than men, whilst there are no differences in dance-confidence self-esteem between men and women over the age of 60. This is interesting because other published studies, for example Robins and Trzesniewski (2005) report that in general men have higher self-esteem than women and Gentile, Grabe, Dolan-Pascoe, Twenge, Wells & Maitino (2009) have found that women have higher self-esteem in only two out of ten areas. Women, for example, are reported to have higher self-esteem than men in behavioral conduct and moral-ethical self-esteem, but lower self-esteem than men in physical appearance, athletic, personal self and self-satisfaction self-esteem. There are no sex differences in self-esteem in academic, social acceptance, family and affect self-esteem. My findings add to these findings and suggest that women, but only women under the age of 60, have higher levels of dance-confidence self-esteem than men.

Now we're beginning to see the effect that age has on dance-confidence self-esteem. The differences between the sexes disappear when people get older. Other researchers have mapped out how general self-esteem changes across the lifespan. See Robins and Trzesniewski (2005) for a review. I'll tell you about that first so that we can compare how dance-confidence self-esteem compares with general self-esteem across the lifespan.

General self-esteem
General self-esteem has been found to change across the lifespan at key developmental periods. The developmental changes in general self-esteem show that for both men and women, general self-esteem is higher for younger teenagers, below the age of 18, than for those in the 18-22 age group. General self-esteem appears not to change between the ages 23 and 49, but it then rises steadily

during the 50s and 60s before it drops for both men and women who are in their 70s and 80s. With this in mind let's have a look at what I found. I'll first tell you about young people, then middle-aged people then older people.

Young People, Self-esteem and Dance

I found that although levels of dance-confidence self-esteem were very high in girls below the age of 16, dance-confidence self-esteem dropped dramatically in girls over the age of 16, to almost the lowest point in a woman's life. But why should this be the case? One way of explaining this is by using the Competency Model. Let's assume that younger and older young-women use social recreational dance for different purposes and this change in purpose reflects a change in their competence at social dance. Let's assume that for girls under sixteen, recreational dance is simply a fun enjoyable activity. Girls of this age might either dance on their own at home, in small groups with other girls or they might attend formal dance classes. All these activities increase their dance confidence because young girls are accomplished and competent at using dance for these purposes. When girls get to about sixteen they often give up formal dance classes. I have asked hundreds of girls about their dance history and of those girls who had stopped dancing most had stopped taking formal dance classes between the ages of 15 and 18. At about the same time young women are likely to start dancing publicly in front of members of the opposite sex (at social parties or discos and so on). When young women start dancing in this way, publicly, they are not accomplished in this use of social recreational dance, especially when they first embark on it and therefore, according to the Competency Model, this would explain why they have lower levels of dance-confidence self-esteem. According to this hypothesis dance confidence should increase as young women become more accomplished and competent at this new use of social recreational dance, and this is indeed what I have found.

The Competency Model can also account for the low levels of dance confidence I have seen in boys under the age of sixteen and in the steady increase in dance-confidence self-esteem that I have seen as boys get older. Boys under the age of sixteen do not enjoy the same advantage of high dance-confidence self-esteem as girls of the same age because they do not, on the whole, use dance in the same way as girls during this period of their development. Young boys are less likely to dance with male friends at home and they are less likely to enrol in formal dance classes than girls.

The Reflective Appraisal Model can also account for the rise in dance confidence in young men in their late teens and early twenties. The Reflective Appraisal Model predicts that if other people respond positively to a person's social dancing then that person will develop a high level of dance-confidence self-esteem. According to Darwin (1872) one function of social dance for young men is as a courtship display. Social recreational dance is therefore part of the sexual selection, or mate-selection, process. According to this view young men dance for two reasons: to display themselves to potential mates and to make themselves stand out from their mate-selection competitors. The way young men dance, and the way they are perceived by women, is related to both their hormonal make-up and their genetic quality. In social dance settings, such as discos and nightclubs, men are clearly being watched and evaluated as potential mates. Evaluation such as this is just the type of social feedback that the Reflective Appraisal Model suggests will have an impact on a person's dance-confidence self-esteem. As men and women use dance more and more as part of the mate-selection process during their late teens and early twenties both sexes will become more competent with this function of dance and both will get used to receiving feedback from other people and this will lead to higher ratings of dance-confidence self-esteem during this period. We're used to being told to dance as if no-one is watching, but the reality, especially when you dance in public, is that we're being watched constantly.

Being watched and judged plays a crucial role in influencing how people feel about themselves. This is especially true for boys who dance, or for boys who think about and reject dance as an active pleasure pursuit. Several researchers have written about the negative feelings young men have about dancing, see, for example, Burt (1995) and Fisher & Shay (2009).

Nadine Holdsworth (2013) writes about a project called Boys Dancing, which, she argues, challenged the normative discourse around gender, boys and dancing and highlights some of the issues that may lead young men to question whether dance is the most appropriate activity for them to engage. The Boys Dancing project was set up to challenge the perception that dance is primarily a female activity (see Risner, 2009) and it gave the opportunity for boys and young men to engage in a range of performance making activities. While the Boys Dancing team found it easy to recruit boys to the project from primary school (5-11 year olds) it was more difficult to recruit older boys from youth centres, who expressed reservations about fears of being called "gay", something that may challenge a young man's sense of masculinity (Holdsworth, 2013). Holdsworth writes, "…in many societies in the West, the male dancing body challenges the very foundations of the masculine ideal and, as such, the male dancer is more often than not connected to a peripheral, failing masculinity or derided as effeminate, 'where "effeminate" is a code word for homosexual', regardless of the male dancer's actual sexuality (Burt, 2007)" (Holdsworth, 2013, p. 170).

This is certainly something I experienced as a young man who danced throughout his teens, an experience from which I recognize the comments of one of the participants of the Boys Dancing Project who said that dancing led some of his peers to "call me queer, they call me gay-boy, they call me gaybo, everything around the lines of homosexual.' (Holdsworth, 2013, p. 176). It is feelings such as these, feelings that perhaps the act of dancing challenges young men's perception of masculinity, that contribute to young men's dance confidence being lower than that of age-matched young women.

Middle-aged People, Self-esteem and Dance

For middle-aged people, I found that dance-confidence self-esteem levels were much higher in women than in men and that dance-confidence self-esteem levels do not change very much between the late-twenties and mid-fifties. One suggestion for the stability of dance-confidence self-esteem levels during this period is that in terms of the mate-selection process, this is the period when people typically marry and have children. According to the UK Office for National Statistics (2005) the average age of first marriage in the UK in 1991 was 27 for men and 25 for women, and the average age of childbearing for women was 29. These data suggest that from their mid-twenties onwards people are moving away from the stage of mate selection and are beginning to settle down with their partners and start a family. If social and recreational dance forms part of the courtship process it is unsurprising that no major age-related changes occur to dance-confidence self-esteem during this time.

However, many first marriages break down and people remarry. The UK Office for National Statistics report the average age of marriage for people who have previously been divorced is 37 for women and 40 for men. So, people who are divorced are likely to enter the mate-selection process again and they may use recreational dance once again as part of this process. What I would be really fascinated to find out is the extent to which divorced people use recreational dance as part of the mate-selection process in later life and also to understand if the dance-confidence self-esteem levels of divorced people who are using dance as part of the mate selection process are different from the levels of people of the same age who are married and no longer using dance as part of the mate-selection process.

The Reflective Appraisals Model would predict higher dance-confidence self-esteem in such divorced people because they will be receiving more feedback on their social dancing than married people who are not engaging in social and recreational dance for the same reason. It would also be really interesting to compare the same thing in people of this age who are happily married and still enjoy social and recreational dancing but without their partners.

Older People, Self-esteem and Dance

In older people, I found that dance-confidence self-esteem changed as a function of both sex and age group. I found that for people in the 56-60 age group, women had higher levels of dance-confidence self-esteem than men. This finding is consistent with similar gender differences observed in all younger age groups. However, I also found that for the first time across the lifespan there was no difference between men and women in dance confidence in the over sixties. This closing of the dance-confidence self-esteem gap has two aspects. First, dance-confidence self-esteem drops measurably for women over the age of 60 and, second, dance-confidence self-esteem increases measurably for men over the age of 60.

The drop in dance-confidence self-esteem seen in women over the age of 60 is consistent with the decline in general self-esteem described above. There are several possible explanations for such an age-related decline in general self-esteem. These explanations are based on situational factors, such as retirement, loss of life partner, declining health and lower socioeconomic status. It is clear to see how such factors might have a negative impact on both global self-esteem and on dance-confidence self-esteem in particular. Loss of a partner, reduced income and serious negative health could all prevent someone from engaging in social and recreational dance and this reduced participation might, according to the Competency Model, make people feel less accomplished at the activity. However, if this were the main explanation for a decline in dance-confidence self-esteem then we should expect to see a similar decline in dance confidence in men as well as women, which we do not see.

Another explanation could be linked to this period being the time when women become post-menopausal. The post-menopausal stage is the time when women can no longer conceive a child. If perceptions of dance ability are related to fertility-based courtship displays then it is unsurprising that a negative change in dance-confidence self-esteem occurs during this stage in a woman's life.

Justine Coupland, from Cardiff University, carried out an ethnographic study of older women (42-74 year olds) who took part in recreational dance classes (Coupland, 2013). Ethnography is the study of people and cultures. An ethnographic study is one where the researchers immerse themselves in a group of people and observe behaviour from the inside, and as part of, the group. This gives the researcher a privileged, intimate, position from which to comment. Coupland describes a conflict in issues of "watchability" in older female recreational dancers. The conflict is centred on the notion that dance is, inherently, a "watched" activity, whether other people do the watching or whether it is a self-watched activity as we dance in front of mirrors. This is in conflict with the idea that, according to some researchers, older age is construed as "unwatchable" (see Woodward, 1991).

In Coupland's study the women chose not to look at themselves in mirrors as they danced and discussed issues of the look of ageing bodies. One of the participants "Sarah" said that she

"...would rather not see a mirror, it will often make me feel more self-conscious, not a positive thing in a dance session." (Coupland, 2013, p. 13)

and "Linda" spoke of her habit of hiding her body in other mirrored places

"I find the length of time in front of a mirror at the hairdressers a bit stressful. I find I am hiding my hands – they look so old!" (op. cit. p. 13).

Coupland's study highlights how older women, and some younger women (let's face it, 42 is not old), deal with and come to terms with dancing in older bodies.

It is not clear why dance-confidence self-esteem levels go up in men over the age of sixty. One suggestion is that the dance-confidence self-esteem of men under the age of 60 is negatively influenced by the high dance-confidence self-esteem of women, to the extent that the men might be intimidated by the women's high confidence. The observed decrease in women's dance-confidence self-esteem may release men from their lack of confidence in their own dance ability: perhaps because they no longer feel so intimidated by high levels of female dance confidence, their own dance confidence rises.

Clearly, we need to do more research to understand the reasons why dance-confidence self-esteem levels increase for men over 60 and decrease for women over 60. Another possible explanation could be due to older people historically being more involved in partner dancing, which might lead to equality of dance-confidence self-esteem scores between the sexes. This could be tested quite easily. If we were able to collect data on dance-confidence self-esteem with the same people repeatedly over many years, it would be interesting to see if male and female dance-confidence self-esteem scores come together when today's young people reach sixty. As there are differences between men and women in dance-confidence self-esteem scores in all age groups up to fifty-five, we only have to wait five years to see if the dance confidence scores of people who are currently fifty-five change as they get older.

We could also examine the dance-confidence scores of younger people who currently engage in partner dance (e.g. Lindy Hop, Ballroom, Latin and so on). If there is no difference in dance-confidence self-esteem scores in younger partner-dancers, then we might conclude that partner dancing is responsible for the equalizing effect on dance confidence. Nevertheless, it is not clear why partner dancing should reduce the dance confidence of women.

Summary

My survey of nearly 14,000 people has shown very clearly that dance-confidence self-esteem changes as a function of sex and age. However, these are not the only factors that will influence someone's dance-confidence self-esteem, factors such as prior dance experience, cultural background, relationship status, attitude to physical exercise and general health status may also contribute.

CHAPTER 6

Dance, Hearts and Well-being

In 2014 a new way of handling health care budgets was introduced in the UK. The Personal Health Budget (PHB) is an amount of money given to certain individuals that enables them to directly purchase the healthcare they need based around an agreed health plan. So, for example, if a person has obesity, and an agreed health plan based around weight reduction, they can either buy "traditional" healthcare, such as paying to see a dietician, or they can buy "different" healthcare, such as paying to join an online diet service, or they can use their personal heath budget to buy "radical" healthcare, such as attending dance lessons (Vogel, 2012). If people are to use PHB's to purchase what might be considered "radical" healthcare, such as dance-based interventions, there needs to be evidence that such interventions are effective, and not a waste of public money. Within this context it is worth noting that Lauren Vogel (2012), in a news article on the use of dance lessons on PHB's reports a case where a personal health budget was shut down because people were spending their budgets on, amongst other things, pole-dancing.

There is clear evidence that engaging in physical activity can lead to extraordinary physical health benefits. For example, regular and moderate physical activity has been shown to reduce the risk of coronary heart disease, and among people with type 2 diabetes physical activity can reduce the risk of related medical complications. The consequences of physical inactivity are also very important as they are related to the incidence of cancer and mortality in some cancer populations (Penedo & Dahn, 2005). With regards to cardiovascular disease a study by Wessel et al. (2004) into the relationship between physical fitness and coronary artery disease, and cardiovascular events, in a sample of 906 women found that lower self-rated scores on physical fitness were associated with higher prevalence of coronary heart disease risk factors (which include risk factors associated with high blood pressure and cholesterol, diabetes and prediabetes) and coronary artery disease, as observed through coronary angiograms which is a test of the main arteries that supply the heart with blood and oxygen. There seems little doubt that physical exercise is good for our heart and well-being.

Measuring Activity – METs

The intensity of different physical activities is measured in units called METs. A MET is an estimate of the metabolic equivalent intensity level of performing a certain physical activity. Different physical activities require a different level of physical intensity. For example, resting (sitting quietly in a chair doing nothing) requires very little physical intensity and so has a MET score of 1. MET scores for different activities are calculated as multiples of the resting MET. So, fast running requires a lot of physical intensity and requires eighteen times more intensity than resting, and therefore has a MET of 18 (the highest MET). All other physical activities come somewhere between 1 and 18. So, the fast chopping of a tree with an axe has a MET of 17, boxing in a ring has a MET of 12, general bike riding has a MET of 8 and sexual activity has a MET somewhere between 1 and 1.5. Wessel et al. (2004) observed a relationship between the MET intensity of physical activity and the risk of major adverse cardiovascular events (such as heart attacks and death) in their sample of women. They report that for every 1 MET increase in the physical activity undertaken there was an 8% decrease in risk of major adverse cardiovascular events approximately four years later.

But what about dancing? Is it a form of physical exercise that provides enough activity to lead to significant changes in health and well-being? MET intensity levels for different types of dance range from 2.5 to 10, which means dancing is more intensive than playing croquet and as intensive as taking part in track and field sports, such as steeplechase and hurdles. MET values for a range of activities are listed in Ainsworth et al. (2000) and examples related to recreational dance are shown on the next page.

In summary, conditioning exercises have MET values between 2.5 - 6, ballroom dancing ranges from 3 – 5.5, spiritual dancing in church has a MET value of 5, ballet, tap and modern are 4.8, traditional national dances ranges from 4.5 – 5.5 and different forms of aerobic dance range from 5 to 10 METs.

Of course, there will be variation depending on how each form of dance is taught by different teachers and engaged with by different dancers.

MET intensities for several forms of dance, and activities related to dance.

Code	METS	Activity	Examples
02100	2.5	Conditioning exercise	Stretching, hatha yoga
03040	3	Dancing	Ballroom, slow (e.g. waltz, foxtrot, slow dancing), samba, tango, 19th C, mambo, chacha.
03025	4.5	Dancing	General, Greek, Middle Eastern, hula, flamenco, belly, swing
03031	4.5	Dancing	Ballroom, fast (disco, folk, square), line dancing, Irish step dancing, polka, contra, country.
03010	4.8	Dancing	Ballet or modern, twist, jazz, tap, jitterbug
20040	5	Religious activities	Praise with dance or run, spiritual dancing in church
03020	5	Dancing	Aerobic, low impact
03030	5.5	Dancing	Ballroom, fast (Taylor Code 125)
03050	5.5	Dancing	Anishinaabe Jingle Dancing or other traditional American Indian dancing.
02090	6	Conditioning exercise	Slimnastics, jazzercise
03015	6.5	Dancing	Aerobic, general
03021	7	Dancing	Aerobic, high impact
03016	8.5	Dancing	Aerobic, step, with 6-8 inch step
03017	10	Dancing	Aerobic, step, with 10-12 inch step

The average METS for dance and dance related activities is 5.8 (excluding hatha yoga).

Key Studies Exploring Dance & Physical Health in Young People

Dance for health (Flores, 1995)

The aim of the project was to see if attending a series of regular dance classes would help to reduce the BMI and resting heart rate of young children. Flores points out that sedentary lifestyles and being overweight are contributors to the development of cardiovascular disease in later life. The link between aerobic fitness, BMI and cardiovascular disease risk factors in adolescents has been reported by Eisenmann et al. (2005). In a study of over 700 children between the ages of 9 and eighteen, Eisenmann et al. found that those who had the highest aerobic fitness and a low BMI showed the best cardiovascular disease risk factor profile, whereas those who had the lowest aerobic fitness and a high BMI showed the worst cardiovascular disease risk factor profile. The profile of cardiovascular disease risk factors was generated by measuring lipids and lipidproteins (triglyceride, high-density lipoprotein cholesterol, total cholesterol, low-density lipoprotein cholesterol); Blood pressure (resting systolic and diastolic and mean arterial pressure); Plasma glucose (fasting plasma glucose concentrations). Eisenmann et al. (2005) also found that within body-size categories (for those people who were either high or low in BMI) aerobic fitness levels were associated with better cardiovascular disease risk factor profiles. In other words, even for those with a high BMI, higher levels of aerobic fitness are associated with reduced cardiovascular disease risk profile.

Within this theoretical context Flores sought to test whether engaging in 36 moderate to high intensity aerobic dance sessions over a 12 week period would lead to a decrease in BMI and an increase in aerobic fitness (as measured by changes in resting heart rate). Flores tested a sample of 81 seventh grade students in the

USA (mean age 12.6 years) using a between groups, repeated measures, design she allocated some of the students to a dance group and the other students to a normal physical activity control group. Flores reports her data separately for boys and girls. For the girls, taking part in 36 dance sessions led to greater positive changes (reductions) in both resting heart rate and BMI than were observed for the girls in the control group. However, no statistically significant changes were observed in the boys who took part in the dance sessions. For girls, at least, increased engagement with dance (at MET intensity 7) might be an appropriate primary prevention activity, which could help reduce cardiovascular disease risk factors.

A "primary prevention" program is the first level of health care, which is designed to promote health and prevent the occurrence of disease. It seems from the study carried out by Flores that dance can play a significant role in promoting healthy behaviours and reducing cardiovascular disease risk factors in young girls.

Differential benefits of dance for girls and boys (Quin et al., 2007)
A similar pattern of findings was found where 348 children aged between 11 and 14 years took part in ten weeks of creative dance. The children took a series of physiological assessments before and after the 10-week programme of dance. Quin et al. found that for the girls in the sample there were significant increases in lung function (forced expiratory volume and functional vital capacity) and in aerobic fitness (20-meter shuttle run test), but there were no significant changes in the scores for the boys. It is worth noting that the girls in the sample were more positive about the dance programme than the boys.

The locus of the sex differences observed in these studies is unclear. We do not know whether the boys engaged with the dance program as fully as the girls, or whether the MET intensity of the classes was insufficient to give the boys a full aerobic workout. It might, conceivably, be the case that boys do not attain health benefits from recreational dance due to differences in their physiological make up, or it might be the case that boys' attitudes to dance somehow interact with the physical benefits of dance.

Dance-based intervention can increase aerobic activity in young men (Adiputra et al., 1996)

It appears to be the case that in at least one study of young men (17-19 year olds), who engaged in 24 dance sessions over an 8-week period, there was a significant increase in aerobic capacity. The study was carried out by Adiputra et al. (1996) using Balinese dance. Balinese dance is a traditional form of dance, used for both religious and artistic expression among the people of Bali, Indonesia. Adiputra et al. divided a cohort of young men into two groups, a dance group and a control (no-dance) group and then took measures of maximum aerobic capacity (VO2max) eight weeks apart. The researchers found that VO2max scores increased significantly for those who took part in the dance sessions and, unsurprisingly, found no change in the no-dance group. This suggests that using a dance-based intervention in a group of young men can increase aerobic capacity.

It is interesting that there are differences in outcome between the studies of Flores (1995), Quin et al. (2007) and Adiputra et al. (1996) with respect to teenage boys' aerobic response to different forms of dance. Whereas Flores, who used Hip Hop dance, and Quin et al. who used creative dance, found no positive effect of dance on young men's aerobic fitness, Adiputra et al.'s use of traditional dance did find a positive effect of dance on aerobic capacity. One of the major differences between these forms of dance is that Balinese dance is both culturally and religiously important and perhaps the importance of the dance form to the participants influenced how they engaged with dance sessions. It is an open question whether asking the boys from Bali to engage in 24 sessions of Hip Hop or creative dance would have led to similar changes in aerobic capacity.

Regardless of the anomalous findings associated with boys, dance and cardiovascular fitness, there is a consistency in the literature which shows that for young people dancing has a positive effect on cardiovascular function and this has the potential to improve health. In addition to the studies already discussed, positive effects of dance on cardiovascular fitness/function have also been reported by Blackman et al. (1988) in a small group of sixteen 14-15 year old girls, by Viscki-Stalec et al. (2007) in a larger sample of two hundred and twenty girls aged between 16-18 year and by Mavradis et al. (2004) in a sample of 6-7 year old children. For a systematic review of this area see Burkhardt & Brennan (2012).

Key Studies Exploring Dance & Physical Health in Older People

All of the studies examined so far have looked at the effect of dance on the cardiovascular system of relatively healthy young people. If some forms of dance satisfy those conditions that are required to lead to healthy changes in the cardiovascular system then dance might help those people who have chronic heart failure or in those with age-related reductions in cardiac function.

It is widely reported that coronary heart disease (CHD) is the primary cause of death for both men and women in the United States of America (National Heart, Lung and Blood Institute) and in the United Kingdom (National Health Service). Although

CHD is more common in men than in women, the risk of developing CHD increases with age for both sexes (see Jousilahti et al. 1999). Engaging in recreational dance has consistently been shown to reduce CHD risk factors in groups of older women.

Dancing can significantly improve cardiorespiratory endurance in adults (Hopkins et al., 1999)
David Hopkins and his colleagues recruited 65 women between the ages of 57 and 77 to take part in low-impact aerobic dance classes. Half the women were assigned to a dance group where they took part in three dance sessions per week for 12 weeks. The other women were told that they could join the dance group in twelve weeks time, and in the meantime, they were to continue with their daily activities as normal. Hopkins et al. therefore had an active dance group and a sedentary waiting group. All the women took part in tests of cardiorespiratory endurance, physical flexibility, physical strength/endurance, body agility, motor control & coordination and balance at the beginning of the study and 12 weeks later.

Over the 12-week period clear differences emerged between the active dance group and the sedentary waiting group. For those women in the active dance group there were positive (and significant) changes in their functional fitness measures. They had improved cardiorespiratory endurance, such that they became faster at walking the half-mile walk test, there was an improvement in their strength and endurance, they became more flexible and agile and they showed improvements in their balance. The sedentary waiting group on the other hand saw significant negative changes in their functional fitness measures. Their cardiorespiratory endurance was reduced by 4%, their body agility reduced by 3% and their motor control/coordination deteriorated by 6%. Hopkins et al. conclude that "Without intervention, it appears that sedentary elderly women continue to decline in functional fitness." (p. 191). This study therefore suggests that low-impact aerobic dance sessions can improve cardiorespiratory endurance in sedentary older women.

Two further studies have shown a link between engagement with recreational dance and cardiopulmonary performance.

A Hong Kong study of low-impact aerobic dance classes (Hui et al., 2009)

Ninety-four Chinese women (and three men) between the ages of 60 and 75, either took part in two low-impact aerobic dance classes per week for 12 weeks or they waited for the same period of time. This study was very similar in terms of design to Hopkins et al. (1999) described above. Hui et al. took measures before and after the 12 weeks of dancing or waiting to dance and they report that for those people who were allocated to the dance group there was a greater positive change in resting heart rate, walking speed, dynamic balance and mobility, lower limb endurance and in perceptions of general health than in the waiting group. They conclude that through social contact with other peers dancing can enhance psychosocial well-being.

Turkish dance based programme (Eyigor et al., 2009)

In the second study, carried out in Turkey by Eyigor et al. (2009), 40 women over the age of 65 were allocated to either a Turkish Folkloric dance-based exercise programme or to a control group. Those women in the control group were asked to continue their normal physical activities. Those women in the Turkish Folkloric dance-based exercise group attended three dance classes per week for eight weeks. Eyigor et al. took a range of measures before and after the eight-week intervention and found that for those in the Turkish Folkloric dance-based exercise group there were similar significant improvements in walking speed, lower limb endurance and in perceptions of general health as were reported by Hui et al. (2009). Those people in the control group showed no such improvements. In fact, for those in the control group there was a significant negative change in their perception of general health.

What's the most effective dance dose?

These studies show a convergence of findings on the effect of different types of recreational dance on cardiorespiratory and general health in older women. It seems that several weeks of regular dancing is good for the heart. However, these studies raise many important questions, such as, how much dance is necessary to lead to significant changes in health? We know that three 50-minute sessions per week over 8 weeks is effective, but what is the minimum dose of dance that we can take and still derive a significant benefit? Di Blasio, De Sanctis, Gallina & Ripari (2009) suggest that a single 90-minute session of Caribbean dance has a sufficient impact on metabolic and cardio circulatory systems to potentially improve health. We clearly need more research before we can prescribe a recommended dance dosage. Some other important questions concern how long the positive effects of dance last and whether men can derive a benefit from engaging in recreational dance too.

Long Term Effects of Dance on Physical Health

None of the studies that we've looked at so far have examined how long the positive effects of dance participation last post treatment. Research in the field of osteoarthritis and a physiotherapy-based individual exercise program has shown that the positive effects of intervention, which are seen at the end of a twelve-week treatment period, decline over time during the next twelve- and twenty-four weeks and the positive effects finally disappear altogether (see van Baar et al, 2001). Similar reductions in the positive effects observed after a 16-week cardiovascular or resistance exercise programme in older adults were seen ten weeks after the completion of the study (see Sforzo et al., 1995).

Within the context of cardiovascular risk and dance Kim et al. (2003) carried out a study which looked at the effect of a health promotion programme (which included a programme of Korean traditional dance movements). Kim et al. examined changes in cardiovascular risk factors, health behaviours and life satisfaction in 21 institutionalised women between the ages of 67 and 89. The women were monitored before and after they took part in the study, which lasted for three months, and again 3 months after the conclusion of the study. Kim et al. report follow up data which shows that for some measures the effect of the health promotion programme persists beyond the termination of the programme.

For cardiovascular risk factors, the women started with a low-to-moderate total risk score of 20. A cardiovascular risk factor score is calculated as a measure based on several factors including age, family history, cholesterol and tryglyceride, systolic blood pressure, obesity (BMI), smoking habits, stress, and exercise habits. This score reduced significantly (to 16.8) immediately after the health promotion programme and remained significantly lower than pre-test three months later (at 18.1). It is not entirely clear why a health promotion programme including elements of Korean traditional dance should lead to a longer-term benefit, beyond the end of the treatment programme. Perhaps it is because the participants in Kim et al.'s study were institutionalised and as such were encouraged to continue with some form of social dance once the formal programme had ended, or perhaps the participants enjoyed the dancing and carried on by themselves.

We cannot know whether the participants stopped dancing at the end of the programme and as such we cannot be certain for how long such effects last in the absence of a dance programme. Most of the older published research on the relationship between CHD risk and dance has been carried out on women. However, there have been two studies on the effects of dance-based exercise on adults with chronic heart failure. Belardinelli et al. (2008) compared waltz dancing with traditional exercise training (exercise bike or treadmill) and a control group on a sample of 109 men and 21 women, and found that after 8 weeks of dance or exercise there were significant improvements in a range of measures associated with cardiopulmonary exercise testing (such as the maximum rate of oxygen consumption, peak heart rate and systolic blood pressure). No such changes were observed in the control group. The finding of Belardinelli et al. suggest that dance is as good a form of exercise for this population as traditional exercise.

Drop-out Rates in Physical Activity

One of the major problems with traditional forms of exercise concerns adherence and drop-out rates. The proverb "The road to hell is paved with good intentions" sums up adherence to exercise classes and drop-out rates rather well. The proverb means that people may have good intentions to do something, but ultimately, they fail to take action. This is seen most starkly in unused gym memberships. In 2011 the UK consumer organisation WHICH? published the findings of research on its website which suggests that people in the UK are wasting about £37,000,000 per year on unused gym memberships. People are clearly motivated to join a gym and attend regularly but, the report argues, despite buying all the gear they soon stop attending. Nevertheless, they let the monthly fees continue to be paid, presumably because they intend to return.

Motivation

A study carried out in Greece by Antonia Kaltsatou and her colleagues compared, among other things, cardiopulmonary exercise testing and motivation in a group of 51 men with chronic heart failure (such as coronary artery disease, hypertension, valvular heart disease or arrhythmia). The men were divided randomly into three groups: Group A was a Greek traditional dance group, Group B was an exercise group and Group C was a sedentary control group. Those men in Groups A and B took part in eight months of exercise and were tested at baseline and again at the end of the exercise programme (8 months later). Kaltsatou et al. (2014) report similar findings to Belardinelli et al. (2008) with regards to peak oxygen consumption, such that following the dance and exercise programme there were significant increases in VO_2Peak but there was no such increase in the control group. In addition, Kaltsatou et al. also observed an increase in self-perceptions of general health in those men who were in either of the dance or exercise groups. This finding is consistent with similar patterns of data observed by both Eyigor et al. (2009) and Hui et al. (2009), who both examined cohorts of female participants.

One of the unique findings of Kaltsatou et al.'s study was that levels of intrinsic motivation were higher following 8 months of Greek traditional dance than they were following 8 months of exercise or being sedentary. Intrinsic motivation is a measure of the evaluated subjective experiences of taking part in the activity and it includes dimensions for enjoyment/interest, effort/importance, perceived competence and pressure/tension.

Kaltsatou et al. observed increases in intrinsic motivation but only in those men who took part in the dance sessions. These men showed a significant increase in a score of total intrinsic motivation and also in increases in the elements of enjoyment/interest, effort/importance and perceived competence. There was no change in pressure/tension. These findings also echo participants attendance rates for the different forms of exercise. The average attendance for the men in the dance group was 96.3%, the average attendance for those in the exercise group was just 91.5%.

Summary

It seems, therefore, that taking part in dance-based activities provides the same physical benefits as regular exercise but that it delivers these benefits in a more efficient way because men feel more motivated to attend the classes over a long period of time. Of course, Kaltsatou et al.'s study used traditional Greek dance with cohorts of older Greek men and it remains to be seen whether similar health-related benefits would be observed in the same male population if other forms of dance were used or whether the use of traditional forms of dance might be as beneficial to older men in other parts of the world.

CHAPTER 7

Watching Dance

Watching modern dance can be confusing. Take, for example, the work of Merce Cunningham. Cunningham (1919 – 2009) was one of the most prominent and prolific choreographers of modern dance in the twentieth century. His work exemplifies why modern dance can be so difficult to understand because it often didn't tell a story with a set of coherent elements. He adopted a process of Creative Independence whereby the separate elements of dance, music and design were created in isolation of the other elements. Sometimes the first time the dancers heard the music and encountered the design features, such as costumes, was on their first public performance. Imagine what it must feel like for a dancer to perform in an unfamiliar physical and auditory environment in front of an audience for the first time. If Cunningham made things difficult for his dancers he certainly didn't make things easy for the audience. He avoided using narrative, theme or character in his pieces, and the order of the movements were sometimes decided by the toss of a coin. Imagine taking a journey where every left or right turn was decided by a coin toss. You'd have no idea where you were going, and no idea if you had ever arrived.

Merce Cunningham isn't the only choreographer to make abstract dance. Many other modern dance makers, such as George Balanchine, Paul Taylor, Martha Graham, Lea Anderson and Wayne McGregor, eschew explicit meaning. Judith Mackrell writes in the introduction to her excellent book Reading Dance "...the only basic skill that's needed for reading dance is a curiosity about the event - a willingness to let the movement play on our senses, to let its rhythms charge up our pulses and to let its pictures range around our imagination" (Mackrell, 1997, p. 1). Perhaps we, as audience members, don't need narrative, structure and a sense of expectation, all we need is to let the experience wash over us like an incoming tide.

Making Sense of our Visual World

However, research in psychology shows us that it is very hard to just let an experience wash over us in a passive way. Humans are active perceivers and we are problem solvers who like to resolve ambiguity. Have a look at these illustrations and think about how they might influence what you see when you watch abstract modern dance.

We actively try to make sense of our visual world and one of the ways we do this is to perceive continuity in what we see. How many lines do you see in Figure 1? Most people see two lines crossing, with one line going from left to right and another going down from top left to bottom right. Although some people see four lines converging in the centre point most people see two and this is an example of the law of good continuation. The idea of this law is that we expect things to have a certain shape and dynamic quality and an expectation that along a path there should be no awkward sudden changes of direction. This is why hardly anyone sees two pointed shapes kissing at their apex.

Figure 1: Law of Good Continuation

Here's another example:

Figure 2: Simplicity

How would you describe Figure 2 in geometric terms? Most people would describe the image as a grey square in front of a black triangle. They imagine, or perceive, the black triangle to continue behind the square. Simplifying the image so that its form continues from one side to the other. We imagine the straight contours of the triangle connecting one side of the image with the other. However, the image might also be perceived as constructed of three separate, but touching, shapes (from left to right, a black trapezoid, a grey square and a black triangle) but we tend, when perceiving, to simplify our environment and this is why we don't imagine the lines of the triangle behind the grey square to be wiggly.

Now, what do you see here?

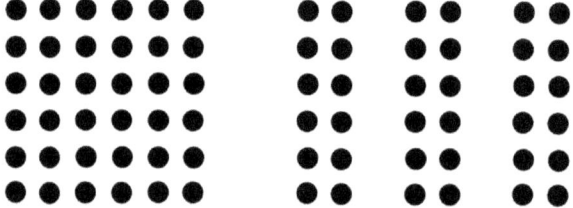

Figure 3: Law of Proximity

Most people report seeing a square of dots on the left and then three columns of pairs of dots on the right. They do this because of the law of proximity, whereby things that are closer together are perceived as belonging together. There is nothing visually connecting the dots to form either of the perceptual square or columns except our expectation that these sets of dots belong together. We don't just see 72 separate dots, we perceive shapes and form based on our expectations and previous knowledge of the world.

These examples, and there are many more, for example the duck/rabbit illusion and the young lady/old lady illusion, demonstrate some of the processes that go on when we perceive either minimal or ambiguous material. They show us how we have an expectation about the dynamic qualities of things we see, such that we expect lines and shapes to have a certain form and travel in predictable directions. When we encounter minimal information, we tend to impose order on what we see by, for example, assuming that things that are close together belong together and these examples show us that one set of visual information can be accurately interpreted in multiple ways. When we apply these processes to watching abstract dance it is clear to see why it is so hard to let the images passively wash over us.

Even when dance doesn't set out to be abstract, and devoid of easily interpretable meaning, the communication of stories through dance is fraught with ambiguity. Consider the ballets Romeo & Juliet and Swan Lake. Even before you watch a performance of either of these you'll probably know what each of the stories is about. You'll know the macro structure of the story (such that in Romeo & Juliet two people from different families fall in forbidden love and tragically die while trying to start a life together) and you'll know some of the details of the micro structure (such that in Swan Lake, Odette is the victim of a spell that has turned her into a swan and she becomes human again at night when she's by the enchanted lake, which was formed by the tears of her mother). Although choreographers have licence to change aspects of a story you'll probably be able to follow either story regardless of which production of Romeo & Juliet or Swan

Lake you see, as long as you know the story first. In terms of psychology your knowledge of the major components or themes of these stories will support your interpretation of what you see on stage, but they won't help you to identify the meaning of every step and gesture.

Read, if you can, the next two paragraphs:

Paragraph 1: I'm sure that as you read this sentence you can understand what I have written. This is probably because I was taught how to spell, and because you were taught how to read English.

Paragraph 2: But waht aobut this? Wehn you raed this, even tgouth the leretts are in the wnorg odrer you can slitl udnersantd waht I hvae wretitn. Radnieg dacne is ambgiuous, jsut lkie redinag tihs setnecne.

Reading dance is, as I've just written, ambiguous, just like paragraph 2. We might understand what is being presented to us but it takes a great deal of active processing. We rely on previous knowledge to comprehend what we see and, together with the elements discussed above, our prior knowledge critically influences the meaning we derive from dance. Several classic studies in the psychology of reading text, and listening to speech, illustrate how our previous knowledge, culture and expectation colour how we make sense and remember stories.

Cultural Influences

Culturally-specific knowledge influences how we make sense of what we read in stories, how much we remember of it and how much we distort what we encounter. For example, Steffensen, Joag-dev & Anderson (1979) asked participants to read two letters that described culturally-specific wedding scenes (Indian or American). The participants were either American or Indian. Steffensen and her colleagues found that when participants read passages associated with their own culture they showed quicker reading times, increased memory for the details of the text, they elaborated on the information given in the letter more and their recall of the passages showed less distortion than when they read the wedding passages associated with a different culture. What this study shows is that our background knowledge influences how we process information and it shows us that this processing is active, such that we don't just experience the world passively, we construct it on an on-going basis. We see the world and we interpret our current experience through a personally constructed lens, a lens constructed from our past experiences.

Context also plays a very important role in comprehension and memory for stories. In dance, the context of a piece can be provided in several ways. It can come from our previous knowledge of the theme of the piece, such that it's based on a popular story, or from the style of the choreographer or from programme notes. With modern dance the context of the piece is seldom provided by the title given to the piece by the dance makers. Research has shown that providing a meaningful title to a piece of text significantly enhances comprehension and memory for the text. What do you make of the following passage?

With the hocked gems financing him our hero bravely defied all scornful laughter that tried to prevent his scheme. "Your eyes deceive," he had said. "An egg, not a table, correctly typifies this unexplored planet." Now three sturdy sisters sought proof. Forging along, sometimes through vast calmness, yet more often over turbulent peaks and valleys, days became weeks as doubters spread fearful rumours about the edge. At last, from nowhere, welcome winged creatures appeared signifying monumental success.

When Dooling & Lachman (1971) asked people to read this passage they found that comprehension was significantly related to whether the passage was given a meaningful title. Some people read the passage without the title (as you have just done) and those people struggled to comprehend it or remember its content. For example, they didn't understand the reference to "three sturdy sisters". However, when the passage was read with its title "Christopher Columbus's Discovery of America" comprehension was vastly improved. Readers were able to segment the text into meaningful units and memory for those units of meaning was higher. For example, the "three sturdy sisters" were conceptualised as La Nina, La Pinta and the Santa Maria, the three ships of his convoy.

What this tells us about watching dance is that our understanding of what we see is influenced by our cultural knowledge and by the context within which we place the work. Choreographers and dance makers can use these features to either enhance our understanding of a piece or to throw us off the scent of meaning in order to give us a more abstract experience.

I went to see an evening of dance called 'Still Current', with Lionel, a TV producer friend. Still Current is a programme of short-ish works by the choreographer Russell Maliphant. After each piece, as the curtain dropped and the house lights came up, one of us would ask the "what did you think of that one?" question. We discussed the charisma of the dancers, their physical presence, the lighting, the relationship between the dancers and, simply, whether we liked it. Once we got beyond describing what we saw and spoke of what we liked and found interesting there were clear differences in our opinions of each piece.

While we had virtually the same visual experience and noticed similar things, such as the lighting, the strength and huge physical presence of the first dancer, the charisma-less second dancer, the master and muse relationship in the final pas de deux (including the momentary gestural power of her hand on the back of his head as she looked into his eyes), we were interested in and liked very different things. While I might have said, "No, I got bored in that one, I didn't like it", Lionel would enthuse about how interesting the piece was, and when I said, "wasn't that enthralling?" Lionel looked at me with a blank expression. It was clear that while Lionel and I had a very similar visual experience, our affective evaluation of what we saw was very different.

A team of researchers from Bangor University in Wales and Radbound University in the Netherlands have examined some of the factors that differentiate people's affective evaluation of dance, and their findings might start to explain why Lionel and I evaluated what we saw so differently. Kirsch, Drommelschmidt and Cross (2013) investigated how different kinds of sensorimotor experience changed spectator's enjoyment of watching dance. Using a classic experimental design they asked people to watch short snippets of dance and then rate how much they liked the dance performance, how complex and how interesting they found it. They then divided the people into three groups and each group had to do something different for a few days. Group 1 had to learn the dances that they'd watched in the clips. Group 2 had to sit and watch the dances in the clips over and over again and listen to the accompanying music and Group 3 had to just listen to the music and not watch or learn the dances. A few days later the people in each group were asked to make the same ratings as before for liking, complexity and interest.

Kirsch et al. found that people's rating of how much they liked and were interested in the dance clips varied as a function of what group they were in, and the largest changes in how much people liked and were interested in the dance clips were seen in those people who had physically learnt the dances. There were no differences in how complex they found the dances. It seems, therefore, that our experience of watching and evaluating dance is influenced, in part, by our sensorimotor experience of dancing. But why should that be the case?

Psychological Categorisation of the Evaluation of Dance

Louise Kirsch and her colleagues suggest four explanations for the relationship between dancing and the affective evaluation of dance: Familiarity, Liking for Complexity, Differences in Sensorimotor Neural Activity and Enhanced Perceptual Fluency. Each of these explanations makes different predictions with regards to how much Lionel and I liked "Still Current".

Familiarity
With regards to familiarity Kirsch et al. (2013) suggest their findings support the idea that increased familiarity is associated with increased liking. However, it isn't just being visually familiar that is important, you also need to have embodied familiarity with an action to really experience an increase in liking for that

movement. However, it is not clear from the study what degree of familiarity is most important in terms of increasing how much you like certain movements. In their study Kirsch et al. asked participants in the dance condition to learn a series of pop-dance sequences using Dance Central on the Xbox Kinect system and it was clips from these sequences that participants had to make affective judgements. This leads to two questions. First, will you get effects of familiarity on dance styles when a different set of movements have been learnt within that style of dance and second, will you get effects of familiarity generalising for steps in one form of dance when you have learnt them in a different form of dance?

For example, if the familiarity effect does not require dancers to embody the exact steps that are being evaluated, just steps from a particular style of dance, then it would be possible to teach people, say, ten hip hop moves and then see an increase in their affective evaluation of a different set of hip hop moves. However, if the familiarity effect does require that dancers embody the exact steps then there should be no improvement in affective evaluation of steps that were not taught and learnt. A similar set of hypotheses can be made about generalisation of familiarity effects across different styles of dance that use similar steps. For example, if the familiarity effect generalises from one form of dance to another then we might expect improvements in affective evaluation of liking for the movements of, say, a pas de basque step that was learnt in one style (e.g. ballet) and evaluated in another (e.g. Scottish Country Dancing).

The differences between Lionel and me, in terms of our affective evaluation of different pieces of Russell Maliphant's "Still Current", might be accounted for in terms of our different levels of familiarity with the movements underlying Russell's choreographic work. Lionel was a dance novice and therefore had very little experience embodying Russell's movement repertoire. I have more experience in dance than Lionel and I might, therefore, have a higher level of familiarity with dancing some of the core movements. However, there is a major caveat to add here. Despite my higher level of physical familiarity with some of the moves I have never learnt to dance any of the repertoire of "Still Current" and, perhaps more importantly, I did not always enjoy "Still Current" more than Lionel. However, if on those occasions where I enjoyed the work more than him then the familiarity effect must have generalised either from the bits and pieces of contemporary dance that I've done over the years or from similar movement patterns that I have learnt in other genres of dance.

Liking for complexity
Kirsch, Drommelschmidt & Cross (2013) point out that the first time their participants had to watch and rate how much they liked the different dance sequences (before any learning, watching or listening took place) there was a relationship between the complexity of the movements and how much they liked it. They found a positive correlation between liking and complexity, such that people preferred more complex movements. What is interesting is that this relationship goes away after several days of dancing and/or watching the dance movements. This suggests that when we first encounter a new style of dance we prefer dance that we perceive as more complex, however, as we become more familiar with the dance style, either through dancing or watching it, complexity becomes less important in influencing how much we like it.

Lionel and I were likely to be making affective evaluations of what we were seeing based on different elements of the dance. As a dance novice Lionel's judgement may well have been positively influenced by the complexity of the movements (such as the intricacy of the footwork, the fast turns and gravity defying lifts) whereas my liking for a particular piece might, according to the findings of Kirsch et al., have been less likely to be influenced by these factors.

Differences in sensorimotor neural activity

Sensorimotor coupling describes the coupling of sensory experiences, such as hearing music, with motor activity, such as body movement. The basic idea is that when we "sense" (hear/see) a piece of music or dance, this triggers neural activity in the brain that supports motor movements. Kirsch et al. discuss the idea that when we have learnt a dance or movement sequence we have set up a neural blueprint that is activated when we either perform or watch the same movement. When we watch a movement that we have already encoded in this way this leads to an implicit desire to embody the movement and this, they suggest, might be associated with greater affective evaluation of that movement. So, because I have more experience than Lionel of both learning different dance styles and of watching dance, my implicit desire to move in response to "Still Current" should have been greater than Lionel's and therefore, if you follow the positive line of Kirsch et al.'s argument, I should have shown more liking for the piece than Lionel.

However, this sensorimotor neural activity hypothesis is, I think, missing something important, and that is the emotional context of the initial learning to dance phase. If seeing a piece of dance that you have already learnt evokes the neural activity that underpins the memory of that dance (and includes the memory of the motor plan) then it might also activate the memory of the emotional state in which it was learnt. This could be positive or negative. For me, learning contemporary dance was an unpleasant ordeal. I loved learning ballet and jazz but couldn't get on with contemporary dance when I learned Graham and Cunningham techniques in my late teens. I felt bony and awkward on the floor. Contractions felt tight and restrictive (as they should) and I could never quite feel the rhythm of the bongo drums as we hopped and rolled across the studio. So, this memory of negative emotions is likely to be stored along with the motor plans and if seeing elements of contemporary dance activate those motor plans they are also likely to activate these negative emotions too, and this may lead me to have less positive evaluations of contemporary dance.

Enhanced perceptual fluency
In the academic study of the philosophy of language there is a debate concerning the relationship between language and thought. Some people believe that thought comes before language (such that we think, and then express that thought through words) and others believe that language determines or influences thought (such that our thinking is constrained by the words and the structure of the language we have available to describe the world). The latter position is called the Linguistic Relativity Principal, and is sometimes known as the Sapir-Whorf Hypothesis (see Harley, 2013). The Sapir-Whorf Hypothesis has a strong form and a weak form. The strong form is that the structure and lexicon of a person's language determines how that person will perceive and conceptualise the world. The weak form is that the structure and lexicon of a person's language influences how that person will perceive and conceptualise the world. Either way, the idea here is that the way we experience the world is based on the sophistication of our language.

Kirsch et al. evoke a similar idea to linguistic relativity when they discuss the influence of perceptual fluency on ratings of how much people like watching certain dance clips. They argue that the more fluently a perceiver can process stimuli the more positive their aesthetic response to that stimulus will become. They argue, after Montero (2012), that in their study physical practice might have led to increased perceptual fluency and this, in turn, may have changed people's affective evaluation. Perceptual fluency is similar to linguistic relativity because both imply a richer way of seeing, dividing and noticing the world and will lead to richer cognitive experiences and thoughts. In relation to dance and perceptual fluency Montero (2012, cited in Kirsch et al. 2013) suggests that dance training can enhance people's perceptual fluency such that they notice qualities about movements that untrained people don't, such as the grace, power and precision of movements.

This reminds me of a story I was told as an undergraduate psychology student. When being taught about the Linguistic Relativity Principal we were told that Eskimos had a richer perceptual experience of a snowy landscape because they had dozens of different words for snow, whereas we, in virtually snow-less southern England, only had a few words for snow. Therefore, the theory went, if I stood with Eddie the Eskimo in snowy Alaska his perception of what was in front of us would be much richer, more detailed and varied than my perception of the same scene, simply because he had more words and a richer linguistic environment with which to process what he was looking at. This is similar to the argument that dancing gives viewers a greater degree of perceptual fluency with which to watch dance.

Unfortunately, the Eskimo and snow example appears to be a myth, a story more enjoyed for its telling than for its factual basis. In an essay called the "Great Eskimo Vocabulary Hoax" Geoffrey Pullum (1991) writes "What a pity the story is unredeemed piffle" (p. 161). He points out how the myth of the enhanced "Eskimo" vocabulary for snow has developed in a series of Chinesewhispers published in newspapers and spoken about in lectures, and he concludes that there is no evidence for the oft-repeated claim that people who live in snowy environments have a richer snow-based vocabulary than people who live in a snow-less environment. We should, I think, also be sceptical of the associated claim that Eskimos have a richer perception of a snowy scene than those of us who live closer to the equator, or, if they do, we should at least conclude that it has nothing to do with their supposed richer vocabulary.

To return to perceptual fluency and the affective evaluation of dance, my dance experience should have equipped me with higher levels of perceptual fluency and, as such, I should have enjoyed watching "Still Current" more than Lionel.

Although Kirsch et al. provide us with four possible explanations for why learning to physically dance a series of dance moves might increase people's liking and interest in these moves, there are additional factors that can explain their observation.

Social and Biological Explanations for Dance Watching

In 2013 Sadler's Wells ripped out some its seats and introduced a mosh pit. This was fairly radical for London's major dance house. The theatre was trying to introduce dance to a new audience as part of its Sampled Season and thought that cheap tickets in a standing room only section in the heart of the auditorium might appeal to newcomers.

In 2014 Tomei and Grivel published the findings of a research study, which implies that Sadler's Wells were right to rip out their seats. Tomei and Grivel asked people to either stand up or sit down as they watched a series of short dance sequences live in a theatre. After watching each sequence people reported how psychologically close they felt to the dancer and the sequence. For example, they rated how much they felt swept away by the dancer, how touched they were by the sequence and how much they wanted to move about. They found that those people who were standing up during the performances felt significantly more psychological closeness towards the dance and the dancers than those people who had been sitting down. They also found that those people who attend dance performances more frequently also felt a greater degree of psychological closeness.

Tomei and Grivel provide two explanations for this finding. The first is a Social Perspective and the second is Biological Perspective.

A social perspective

From a social perspective, Tomei & Grivel (2014) suggest that the act of standing up is one of the first social cues for connecting with other people. As they put it, "standing up in social situations embodies the anticipation of connecting to others" (p. 44). It is worth noting that the participants in their study who stood up for each performance sat down between each performance to make their ratings. Therefore, they stood up directly before each performance. It is unclear whether the same effect would be observed if the participants remained standing for the full duration of the study.

A biological perspective

From a biological perspective, Tomei and Grivel suggest that the increase in psychological closeness might be due to increased heart rate. They found in a second study that heart rate varies as a function of whether someone is sitting or standing, such that higher heart rates were recorded in people who were standing up. They link increased heart rate to social closeness through the fascinating findings of White, Fishbein & Rutstein (1981), who wrote a paper called "Passionate Love and the Misattribution of Arousal". In essence, White et al. showed videos of women, who were either presented to be attractive or unattractive, to men and the men simply had to rate how attractive the women were.

Attractiveness rating were calculated by asking the men to rate the women on the following dimensions: likeable, sympathetic, sincere, shallow, irritating, generous, dependable, considerate, conceited, sexually warm, perceptive, humorous and exciting. However, before the men watched the videos they had to do some exercise. Half of them had to run on the spot for 2 minutes and the other half just had to run on the spot for 15 seconds. This led to differences in physiological arousal, such that half the men were physiologically aroused (with an increased heart rate) and the other half was not (low heart rate). White et al. found that ratings of attractiveness varied systematically with physiological arousal. They found that men in the high heart rate group rated the attractive women as more attractive than men in the low heart rate group, and they rated the less attractive women as less attractive than the men in the low heart rate group.

Summary

So, it seems that heart rate may also be a critical component in influencing how much people feel connected to a dance performance. Taking the findings of Tomei & Grivel (2014) and White et al. (1981) together it seems that Sadler's Wells were playing with fire by removing some of the seats. Having people stand up as they watch the performance might have led to an increase in physiological arousal (increased heart rate) in the spectators and this may amplify their reactions to the dance pieces, such that for those pieces they enjoyed they should really enjoy and for those pieces they didn't enjoy they may have left the theatre with an extreme adverse reaction. A risk-averse theatre manager might choose to leave the seats in place. A courageous one will rip 'em out.

CHAPTER 8

Memory

In the late 1980's I was working as a professional dancer in a London theatre with actors and musicians. At the beginning of the rehearsal process the musicians were given the musical score on sheets of paper. They learnt it with a musical director, who conducted them in the live shows. The actors were given the script, which they read repeatedly and learnt until they were "off the book", meaning that they'd memorised the words. They could keep the script with them during the rehearsal process, in which they made margin notes, and the script was followed by the stage manager during the live shows who would shout out lines whenever actors forgot what they were meant to say. The dancers didn't have the luxury of having the dance routines written down. Rather, a choreographer would teach us long complicated routines, which we had to repeat until they were remembered. Video recording was very expensive in the 1980's so we, as dancers, couldn't use video as an aide-memoir. During the live shows the choreographer wasn't present, either to keep us in time or to remind us of the moves. So how did we do it? How do dancers learn and remember dance routines?

Cognitive Psychology

Being able to learn something relies on you being able to remember it. This sounds obvious. Cognitive psychologists are interested in the processes of learning and memory. They address questions such as, how do people learn? How much can a person learn at any one time? Is what you learn an exact copy of what you've been taught? How does what you already know influence what you learn. What is the best way to test what someone has learnt? Addressing questions like these helps us to understand how dancers learn choreography and it also helps us to understand where difficulties in learning and memory arise. Throughout this chapter I will try to address these questions in terms of academic theory and experimental findings. To begin, I will give a brief introduction to human memory and then I will relate this to trying to understand how dancers learn choreography.

Understanding Memory

To start to get an idea of how memory might be organised in the human cognitive system it's useful to think about a few questions. Is "memory" just one big place where we put everything or is memory organised in different ways? How many different types of memory are there? Why is it possible to remember, in great detail, things that happened ten years ago but yet it is possible to forget things that happened five minutes ago? Do we remember everything? Why are some things easier to remember than others?

Let's start with a bit of memory-theory. Researchers, such as Atkinson & Shiffrin, (1968) and Baddeley (1986), suggest there are two fundamental parts to our memory system. There is long-term memory (LTM) and short-term memory (STM). Short-term memory is also known as Working Memory (WM). LTM is the place we store everything we know. It holds memories from our childhood, the rules of the languages we speak and our autobiographical information. Information in LTM can be stored for a lifetime. You might think of it as a permanent store. STM, on the other hand, is a temporary storage space. It's the space where we process new information. It can only hold a small amount of information and that information can only be held for a very short period of time.

When you learn a dance routine you might have a sense of the difference between STM and LTM. Information can travel between STM and LTM and the process of learning involves getting information from STM into LTM and then integrating it with things you already know (and which are stored in LTM). Information in STM doesn't last very long. It either fades away or new information pushes it out. One way to get information from STM into LTM is to rehearse it, to go over it again and again before it gets lost.

Chunking

The amount of information we can hold at any one time in STM and learn is very limited. Experiments have shown that we can only hold on to between five and nine chunks of information at any one time (Miller, 1956). If we try to hold on to more than that then we run the risk of not being able to learn more than a couple of chunks of information. So, overloading STM can have a detrimental effect on learning. The size of a chunk of information varies depending on your previous knowledge. If you can package together lots of little bits of information into one meaningful chunk then you are able to process it easier than if you are trying to process all the constituent parts of knowledge separately. Let's have a dance-based example.

Imagine you have to learn this sequence of dance steps. How many chunks of information do you have to learn?

Face front.
Start with your feet together.
1. Make a sideways step to the right with your right foot.
2. Make a sideways step to the right with your left foot, crossing your left foot in front of your right.
3. Make a sideways step to the right with your right foot.
4. Tap your left foot on the floor next to your right foot.
5. Make a sideways step to the left with your left foot.
6. Make a sideways step to the left with your right foot, crossing your right foot in front of your left.
7. Make a sideways step to the left with your left foot.
8. Tap your right foot on the floor next to your left foot.

If you have no knowledge of dance then this small sequence might overload your STM with counts and instructions and you might, therefore, find it really hard to learn. However, if you've ever danced a grapevine combination, for example in Line Dancing, then this whole sequence of steps will be recognised as a *grapevine* and will take up just one chunk of information in STM. It will be encoded as "doing the grapevine" and it will leave lots of space in your STM to think about (and learn) other things.

A small-scale research study by Carvalheiro & Rodrigues (2009) confirms this in an experimental setting. They were interested in how much dance-based information (motor actions and movement phrases) people could remember in a short period of time. Based on the two assumptions that adults can generally remember more information than children and people with experience in a particular field can remember more than novices, they constructed a design that had children and adults learning sequences of dance moves. Half of each of the adults (university students) and children (mean age 11) were experienced dancers and the other half was novices. In general, they found, regardless of age, that those with experience in dance remembered more motor actions and movement phrases in the correct order than those who were dance-novices. They also found that within ability levels older people remembered more than younger people but that there was no difference in levels of recall between young dancers and older novices. These findings confirm experimentally that the amount of dance movement phrases people can remember, in the correct order, is influenced by the age and the experience of the person dancing.

Familiarity

Familiarity with learning material (dance in the case of Carvalheiro & Rodrigues, 2009) is important to learning because it eases the transition from STM to LTM. That's why it's easier to learn reconfigurations of things we already know, than it is to learn sets of totally new information. For example, it would be easier to learn the words and their order in List 1 than it would be to learn the words and their order in List 2.

List 1	List 2
Table	Spilorg
Sari	Ripsi
Baboon	Boonab
Mountain	Mungtrit
Wicket	Wockun
Friday	Fadyin
Button	Noobot

The words in List 1 are real words and when you think of them they all have a meaning and a sound and they will be fairly familiar to you, whereas the "words" in List 2 are just made up letter strings. They don't have any meaning, although you might try to give them a meaning as you sound them out, and they will be unfamiliar to you. Studies have shown that lists of familiar known words are easier to learn and remember than either lists of unfamiliar known words or non-words. This shows the contribution of LTM in our ability to learn through STM (see Hulme, Roodenrys, Schweickert, Brown, Martin & Stuart, 1997).

The familiarity effect can also be seen when we try to learn dance routines in different styles of movement. We might be more or less familiar with the "language" of different dance styles. For example, I've done a bit of ballet and am fairly familiar with the ballet language, however, there are some forms of hip hop (such as waacking) with which I am less familiar. Therefore, according to the familiarity theory of STM and LTM, I should find it easier to learn a new sequence of ballet steps than I would a new sequence of waacking movements. Therefore, one way to approach learning a new style of dance is to understand that you need to become familiar with the physical "language" of that dance style, and as you do, then your ability to learn and perform sequences of movements should improve. Don't be disheartened if you try to learn a new style of dance and you find it difficult to pick up sequences of movements, this is understandable, even if you are a proficient dancer in a different style.

Recall

Learning new material is not the only important aspect of memory. The other important aspect is recall, or finding and retrieving information from LTM. Our memory system would be useless if we could store information and not retrieve it, but this is what happens a lot of the time. Memory researchers, such as Shallice and Warrington (1970), have learnt a great deal from studying people with amnesia (memory loss). They have found that it is possible to lose access to your LTM but retain access to your STM. This would mean that people are able to function in the here and now, they would be able to function and remember things in the very short term but they wouldn't be able to recall information from their past. On the other hand, there are people who can access their LTM but who have limited access to their STM. These people are able to recall an amazing amount of information from, say, their childhood, but they wouldn't be able to remember what they had for lunch yesterday. Memory loss is also common in older people and while older people may be able to learn and practice dance moves they learnt when they were younger they might find it more difficult to learn new combinations of dance steps as they get older.

Dancers' Cognition

Dancers in commercial shows have to learn choreography very quickly. A choreographer will demonstrate the moves and the dancers will have to learn them simply by watching and repeating what is done, taking correction either from the choreographer or an assistant choreographer. During this process dancers are unlikely to write down what they have to learn and as such they are relying on their visual memory and on what they have learnt in the past.

Dancers have to learn a huge amount of information. When dancers are preparing to dance in a performance they have to learn and remember dozens of different bits and types of information, for example, there are the entrances and exits, the patterns laid out on the floor, their position relative to other dancers. Then there are the movements of the body and these can be thought of as the gross motor movements, such as the big movements of the body, arms, legs etc., and there are also the fine motor movements. The fine motor movements are things like the movement of fingers and the hands, the facial expression, different attitudes within the body and tensions held and released across the body. Dancers have to learn and remember the emotional content of a piece and typically are expected to project all this while taking on the personality of a character. As you can see, there are layers of subtlety within the movements and dancers are expected to learn and remember them all without the aid of a written score. This is not to say that all these elements of dance cannot be written down, they can.

Movement notation
There are several forms of movement notation, the most notable of which are Benesh Movement Notation (see Causley, 1980) and Laban Movement Notation (see Hutchinson Guest, 2005). However, it takes a high degree of skill and time to both write and read dance notation and so it tends not to be widely used in rehearsal rooms where a whole show might need to be choreographed and learnt in a matter of a few weeks.

Fine and Gross Movement Memory

So, there are gross movements and fine-motor movements, and there's the emotional communication through the body. Dancers also have to learn how to interact with other people on the stage, so their movements are co-ordinated in space and time. In addition, dancers have to learn the music. Now of course when dancers perform there is no conductor keeping them in time, sometimes there is music, sometimes there is no music but generally what dancers have to do is to remember all these things. Earlier on I wrote that dancers have to remember dozens of things. Actually, it is more accurate to say that Dancers have to remember thousands of bits of information. From a memory perspective, this really is quite a remarkable feat and it is, arguably, much greater than the memory feat of the actor or the musician, because they have the books, or they have a conductor and the notes in front of them.

What you learn as a dancer is not always an exact copy of what you've been taught. Sometimes a choreographer will expect dancers to reproduce exactly what they demonstrate but this is not always the case. During rehearsal dancers might use a technique called "marking", which means to reproduce a smaller, more economical approximation of the movements. They might do this to conserve physical energy but it also has an impact on their ability to learn, express and remember other movements too.

What you already know has a substantial impact on how you learn new information. Your learning is effected by your familiarity with the to-be-learned material, how frequently you've encountered it in the past and by the meaning you give to what you're learning. So, if a new set of movements is made up from unfamiliar combinations that you have rarely encountered that have no meaning for you then you will find it much harder to learn and remember the movement sequence.

Techniques for Dance Memory

While it is clear that some people have better memory for motor movements than others there are also a number of techniques that people can use to help improve their memory and learning of sequences of dance. These techniques are *marking* and *sleep*.

Marking

Marking is a technique used by dancers to help them learn and rehearse sequences of dance movements. Marking is the opposite of dancing something "full out". Dancing "full out" means doing everything exactly as it should be done; jumping to full height, turning as many pirouettes as required and fully extending limbs. So, if marking is the opposite of dancing full out in means doing the movements in a smaller, energy conserving way. It might also mean replacing one set of movements with another set of movements. For example, in a ballet class a teacher might demonstrate a set of exercises for the feet and legs and dancers will often model the exercise with their hands, before they do it full out with their feet and legs. In other words, the dancers are marking the exercise with their hands.

I've come across several explanations for why marking might be a useful technique. Some of these explanations are intuitive; some ludicrous; some are usefully interesting. For example, marking might be seen as a way of conserving physical energy. After all, dancing full out can be physically exhausting and so marking might, intuitively, be one way that a dancer can get through several hours of rehearsal without becoming physically worn out. Now for the ludicrous explanation. Several years ago, I was in a ballet class for dance teachers and the instructor was telling us about the benefits of marking footwork with our hands. He said that students will remember sequences of steps better if they model them with their hands because their hands are closer to their brain than are their feet. As far as I am aware there is no scientific evidence for the claim that learning success is a function of proximity from the brain (if this was true then it would be easier to learn a sequence of shoulder movements than a sequence of knee movements, and I don't think that's the case.)

A usefully interesting account of the cognitive benefits of marking, one supported by experimental evidence, has been given by Warburton, Wilson, Lynch and Cuykendall (2013), which appears to show that using a marking technique during rehearsal can help people to learn and perform expressive movement qualities.

Warburton et al. asked a group of advanced ballet trainees to learn two 64-count ballet sequences. In addition to learning the sequence of steps the dancers had to learn to embellish the sequence with certain qualities, such as slashing, punching, pressing, wringing, dabbing, gliding, flicking and floating. For example, the movements "single pique passé turn" should be danced with a floating quality. After learning the sequences the dancers were given rehearsal time. However, for one of the sequences they were instructed to rehearse by dancing the piece full out and for the other sequence they were instructed to rehearse using a marking technique. A few days later the dancers were tested on their memory for the sequences, such that they had to dance them full out in a studio during which time they were video recorded. Warburton et al. then analysed the videos and they scored the dances for the accurate reproduction of both the steps and the expression of the qualities.

Scoring the accuracy of the steps in each dance sequence was straightforward, as the dancer either performed the correct step or they didn't. However, the scoring of the movement qualities was less concrete. Judges used Laban Movement Analysis (Laban, 1947) to describe each movement quality in terms of weight, time and spatial intention. Each of these dimensions was scored using a binary method, such that weight was considered either light (feather like) or strong (powerful), time was considered either sudden (quick & urgent) or sustained (stretching or prolonged) and spatial intention was either direct (single-focused) or indirect (multi-focused). Qualities were scored by two assessors.

Warburton et al. found a higher level of dance accuracy, in terms of the correct performance of movement qualities, for those dance sequences rehearsed using the marking technique compared to the performance of the dance sequences rehearsed full out. In one of their analyses they report 91.4% accuracy for the performance of the qualities that had been rehearsed using the marking technique and only 76.8% accuracy for the performance of the qualities that had been rehearsed using the full out technique.

Therefore, it seems that marking has a beneficial effect on memory and performance. But how do they account for such a finding, it's surely not do with practicing with muscle groups closer to the brain. No.

Warburton et al. suggest this finding supports an embodied-cognitive-load theory. There are only so many things that we can attend to or think about at the same time. This is known as cognitive load. If, for example, we use all of our cognitive resources on Task A then there will be nothing left for us to work on Task B. Because learning and performing multi-layered dance routines uses lots of different cognitive resources it becomes inevitable that we will quickly reach our cognitive load and therefore not be able to learn or practice new material. Marking, according to Warburton et al., is thought to use less cognitive resources than dancing full out, which they suggest means we are able to use spare resources to better learn and perform the emotional qualities (the additional layers) of dance routines. The theory is called embodied-cognitive-load because it is thought that using our body takes up precious cognitive resources and the more economically we move it the more resources we'll have for other cognitive tasks.

Sleep

Let's imagine you're faced with a decision. You have to remember a dance routine that you've just learnt. What should you do, physically rehearse it over and over again, or go to bed? The findings of many experimental studies suggest that you should go to bed, and sleep.

According to Walker (2005) the process of memory formation, following learning, involves two distinct stages. The first is a process of stabilization, whereby a memory is maintained at the level at which it was learnt. During stabilization memories become resistant to interference and memory loss. The second is a process of enhancement, whereby the memory becomes more fully integrated with previous knowledge. Enhancement can occur in the absence of physical or conscious rehearsal. Although the process of stabilization is not dependent on sleep, enhancement is. What's the evidence for this?

When we dance we make both *fine* motor movements, such as small movements with our fingers and hands, and *gross* motor movements, such as larger movements with our arms, legs and body. Research has shown that sleep leads to enhanced performance in both fine motor movements (e.g. Fischer, Hallschmid, Elsner & Born, 2002; Plihal & Born, 1997; Walker, Brakefield, Morgan, Hobson & Stickgold, 2002) and gross motor movements (e.g. Genzel, Quack, Jager, Konrad, Steiger & Dresler, 2012; Kempler & Richmond, 2012).

Fine motor movements

Walker et al. (2002) trained 62 right-handed people on a sequential finger-tapping task. The task was to press five numbers on a computer keyboard in a particular order (4-1-3-2-4) as many times as possible in 30 seconds. The task had to be completed using the fingers on the left hand and was measured for both speed and accuracy. The participants were split into groups and learnt the finger-tapping task at different times of the day. Retesting of the learnt skill was carried out 12 and 24 hours after initial learning. Retesting after 12 hours is called Retest 1 (RT1) and retesting after 24 hours is called Retest 2 (RT2). The groups varied in terms of when they slept, such that some of them slept after initial training and before RT1 and others slept between RT1 and RT2.

If learning of the finger-tapping task (stabilization) has taken place then performance of the task at RT1 and RT2 should be the same as it was at the end of the learning phase. If stabilization of learning has not occurred then performance at RT1 and RT2 should be worse than at the end of the learning phase. If enhancement of the finger-tapping task has taken place then performance should improve at either RT1 or RT2, depending on when participants slept.

Walker et al. (2002) found that practice improved performance in all groups of participants during the initial training phase, and that performance levels remained at least at the same level at both RT1 and RT2. This suggests that stabilization had occurred and it confirms the importance of practice on learning fine motor skills.

They also found that subsequent improvements in performance were related to when participants slept. They found that participants who were trained in the morning showed no significant improvements in performance at RT1, when they had been awake all day, but they showed an 18.9% significant improvement in performance at RT2, after they had had a night of sleep. They observed a similar finding for those who had been trained at night, such that they showed a 20.5% significant improvement in performance at RT1, following a night of sleep, but no further improvement in performance when tested at RT2 after they had been awake all day. These patterns of findings suggest that enhancement in learning and memory occurred while the participants were asleep but not while they were awake.

Walker et al. conclude that enhancements in fine motor skill performance are due to changes in brain state that occur while we sleep. The exact changes in brain state are currently unknown but may be related to stage 2 deep non-rapid eye movement (NRAM) sleep. Walker et al. suggest that 52% of the variance in sleep-induced improvements in learning can be explained by the amount of stage 2 NRAM sleep (see also Smith and MacNeill, 1994, who found that disruption of stage 2 NRAM sleep can impair retention of fine motor movements).

Enhanced performance of fine motor movements is associated with periods of sleep. It seems that while we sleep our learnt movement patterns become consolidated with previous knowledge and we are able to execute them faster, without loss of accuracy.

Gross motor movement: dance studies
Genzel, Quack, Jager, Konrad, Steiger & Dresler (2012) examined the effect of sleep on men's ability to learn gross complex motor sequences. Using the DanceStage game on Playstation 2 thirty-six men had to learn a dance routine to one particular song. The way the DanceStage game works is that you stand in the middle of a mat and move your feet to touch areas on the mat either in front, to the side or behind you as instructed to do so by the presence of arrows on a screen. The arrows and the movements correspond with the timing (beat) of the music. The game records

how well the dancer is doing in terms of whether they are hitting the correct positions with their feet and whether they are hitting their marks on the correct beat. It also provides constant visual feedback for the dancer.

Participants danced one particular dance routine nine times over a 24-hour period. Three times as part of the Learning Phase (LP), three times as part of the first Retest (RT1) and three times as part of the second Retest (RT2). There was a 12-hour break between LP and RT1 and also between RT1 and RT2. Half of the participants were awake between LP and RT1 and asleep between RT1 and RT2 and the other half slept between LP and RT1 and were awake between RT1 and RT2. Participants were asked not to practice on DanceStage between testing sessions. In addition, participants learnt new dance routines immediately after RT1 and RT2.

Genzel et al. (2012) wanted to see if sleep would enhance the learning of complex motor sequences and also whether sleep would enhance the transfer of motor skills learning from one dance to another. They found that performance on the main task did seem to vary as a function of when people slept. There were a number of key comparisons made both between the two groups (Sleep-first group and the Awake-first group) and within the two groups as a function of time of testing.

First, they showed there were no between group differences in LP scores (so both groups learnt the dances equally well during the learning phase) but there were differences between the groups in RT1, such that the Sleep-first group scores were significantly higher than the Awake-first group. This suggested to Genzel et al. (2012) that sleeping after initial learning helps to consolidate the information and shows its effect on the improved performance of the task. This is similar to the Walker's (2005) concept of enhancement. There was no difference in RT2 scores between the two groups. The performance of both groups improved between RT1 and RT2, showing traditional practice effects.

Second, participants' ability to learn the additional new dances immediately after RT1 and RT2 was measured and while performance improved from LP to RT1 there was no change as a function of sleep.

Based on these findings Genzel et al. (2012) conclude that sequence learning per se benefits from sleep but transfer of learning occurs independently of sleep. They explain their results within a schema framework. A schema is, at a fundamental level, a knowledge structure. Initial learning is thought to involve the process of basic schema building. In other words, when we are learning a new set of movement skills we need to build a knowledge structure (schema) to support that new set of movement skills. Transfer of learning, on the other hand, is based on the integration of new information into an existing knowledge structure or schema. Sleep may be important in the building of new knowledge structures and it is such structures that underpin our ability to learn and remember information.

Considering the findings of Genzel et al. (2012) in light of the findings of Walker et al. (2002) it seems to be the case that schema building must, at least, be a two-stage process, where the first stage is not sleep based and the second stage is. So, initial learning of a new motor skill leads to the development of a simple schema that, approximately, stands alone. This can be seen as similar to the process of stabilization as described by Walker (2005) and is not influenced by sleep. The second stage is that of enhancement, where the stand-alone schema becomes increasingly integrated into a person's knowledge structure, as if branches of several schema become entwined. This process is sleep dependent.

Genzel et al.'s (2012) transfer stage, where the learning of a new motor movement pattern is supported by previous knowledge, involves different processes to the enhancement stage and it is not sleep dependent. Transfer possibly involves implicit learning, which is learning that precedes both unintentionally and unconsciously (Shanks, 2005).

Consolidation

From a neurobiological perspective, it is thought that consolidation in memory is facilitated by communication between the hippocampus and the neocortex (see McClelland, McNaughton & O'Reilly, 1995). The hippocampus is an area of the brain that can be thought of as short-term memory or working memory system (see Baddeley, 1986), such that it is a place where we process new information, or information we are currently thinking about. New information needs to be integrated into our long-term memory knowledge structures, and it is thought these are accommodated within the neocortex, the outer layer, of the brain (Tse, Langston, Kakeyama, Bethus, Spooner, Wood, Witter & Morris, 2007; Tse, Takeuchi, Kakayama, Kajii, Okuno, Tohyama, Bito, & Morris, 2011; van Kesteren, Fernandez, Norris & Hermans, 2010; van Kesteren, Rijpkema, Ruiter & Fernandez, 2010). Although information fades away from short-term or working memory quite quickly consolidation into long-term memory can, in some cases, take several years.

Summary

A person's ability to learn, remember and perform dance routines will depend on several factors, such as their knowledge of similar dance styles, their age, their ability to mark movement sequences during rehearsal and when they slept.

CHAPTER 9

Dancers and Dizziness

I remember having a conversation with a circus performer in the 1970's. He told me that, in his experience, one of the major differences between dancers and trapeze artists was that dancers made their tricks look easy while trapeze artists made their tricks look hard. He had a point. Imagine a high-wire act. The audience look up, as they watch someone stand on a platform at the end of a taut horizontal wire, they usually gasp. The performer waits, to build tension, and the audience hold their breath. There's an expectation that the performer will balance on the wire, and then, suddenly, lose it, fail and fall. There might be a safety net, there will almost certainly be a safety harness. This is a difficult balancing trick. The set-up, the apparatus and the suspense tells the audience that what they are about to see is difficult and dangerous. The performer will test the wire with one sliding foot after another and slowly walk from one platform to another, and against all the odds and expectations, will retain their balance even after a mid-rope critical-looking wobble. The performer will, eventually, get to the other side of the rope and the audience will cheer with relief and delight. After all, who would have thought that a trapeze artist would be able to retain their balance under conditions such as these? Phew.

Dancers, on the other hand, are much more low-key. They spin, whip and snap their heads as they turn, jump and balance on the tips of their toes. They throw themselves off balance at break-neck speeds, roll on the floor and recover before extending themselves in anatomically impossible lines as they reach to pluck an imaginary flower from the floor of the stage. The simplicity of the pluck is the benchmark for the ease with which they've done everything that went before. It's so easy, light and automatic. It is this ease which separates them from trapeze artists.

Although dancers and trapeze artists deliver their feats in different ways, they both perform amazing acts of balance and their tricks often require them to overcome what might be perceived as the inevitable consequences of disorientation and dizziness. Spinning in the air the trapeze artist needs to count her turns and be ready to be caught by her swinging partner. She needs to know which way is up. Likewise, a dancer at the end of 32 fouette turns needs to stop, usually facing the audience, and then control the visual perception of the world as it rapidly orbits her visual field.

So how do they do it? To have an understanding of how dancers achieve mastery of their balance and control of their spinning spatial orientation we'll look at the basic mechanisms of the human balance system, we'll look at those elements of dancer training which might help to develop better balance and control, we'll look at the scientific evidence which has examined the hypothesis that dancers have better balance than non-dancers, and we'll finish off by looking at some theories of why dancing might change the way we experience the world.

Understanding Dizziness

Dizziness is commonly thought of as a sense of unsteadiness and of lack of balance and it is one of the sensations associated with vertigo. Vertigo is a perception of movement or whirling, either of you or of surrounding objects. Vertigo includes a feeling of rotation or spinning. Disequilibrium describes a perceptual state of spatial disorientation and is also characterized by feelings of unsteadiness and imbalance.

Feelings of vertigo are entirely normal temporary states that anyone can feel. Longer term and more severe feelings of vertigo can also be experienced due to pathological abnormalities and illness. In this chapter I will confine my discussion to vertigo as a normal temporary state and I will not discuss vertigo as a consequence of an acquired pathological condition. (Interested readers might look at www.vestibular.org for a description of the many pathological causes of vertigo).

The Anatomy of Vertigo

We retain balance and feel steady on our feet because of a combination of signals that are processed by the brain. These signals come from three places: our ears, our body and our eyes. More specifically, the signals that go to the brain from our ears originate in the vestibular system. The signals that go to the brain from our body are influenced by our sense of proprioception and the relationship between our eyes and brain in terms of keeping us on balance needs to be understood in terms of the Vestibular-Ocular Reflex (VOR). In addition to signals coming from these sources to the brain (brainstem), feedback messages are also sent from the brain to the eyes to help maintain steady vision and to the muscles and joints to help posture and maintain balance.

The vestibular system

The vestibular system is situated just inside the ear. We have two vestibular systems, one on the right and one on the left of the head.

The vestibular system is made up of several components. There are three semi-circular canals. These are the anterior, posterior and horizontal canals. The canals contain hairs and fluid. The fluid is called endolymph and as you move your head the fluid moves. As the fluid moves it washes over the small hairs and the movement of these small hairs sends signals to the brain to tell it that the head is moving, and they provide information on the angle and orientation of the head. I imagine the relationship between endolymph and the hairs to be like that of sea weed moving backwards and forwards with the movement of the sea.

Within the vestibular system there is also a component called the Otolith Organ and this is made up of the utricle and the saccule. The utricle and the saccule are sensitive to gravity and linear acceleration. The utricle is sensitive to changes in horizontal movement (side to side and forward and backwards) and the saccule provides information about vertical acceleration (up and down). Together the signals from the vestibular system are sent along the vestibular nerve and this provides information on the angle, orientation and movement of the head.

Vestibular-ocular reflex

There's a technique used in some horror movies to make us feel unsteady and on edge. Imagine the scene, a group of people are walking through a forest at dusk. One of them is holding a hand-held camera and as they walk we see the trees shudder and the angles change quickly from side to side and up and down. As we watch their video recording we get the sensation that we are looking at the world through their eyes and we perceive their (nervous) movement through the movement of the things they are looking at. It can be uncomfortable to watch, not least because we know something awful is inevitably about to be discovered.

In the real world, visual perception doesn't work like that. When we look at a stationary object and make small movements with our head the object doesn't appear to move. The image remains fixed despite our movement. (Try looking at the nose of a person sitting near you (or a light switch) and then move your head from side to side. Their nose shouldn't move, and neither should the light switch). The process responsible for keeping images stable as we move is the Vestibular-ocular Reflex (VOR). In essence, the VOR describes a process whereby the brain sends signals to the eyes to compensate for movement.

When the brain detects movement of the head it sends signals to the muscles on either side of the eyes, which moves them in the opposite direction to that of the movement of the head so that images look still and stable. What's amazing about the VOR is that all this happens automatically and you don't have to be looking at anything in particular for the reflex to kick in. For example, the VOR operates even when you have your eyes closed and when it's dark. Together with the vestibular system the VOR helps people to maintain balance.

Proprioception
Close your eyes and put the middle finger of your left hand on your nose, and then place your left hand on your right knee. Then open your eyes again!

Doing these simple movements demonstrates our sense of where our different body parts are in relation to other parts of our body. Even without looking we can move the tip of one finger to the tip of our nose and then we can locate our left and right legs and move our hand to touch one of our knees. For this to happen successfully we have to be aware of our current body position and of how this changes over time. This is called proprioception. The process of proprioception involves sensory receptors in our muscles, joints, tendons and skin sending messages to the brain. Proprioception plays an important role in balance.

A sense of movement, or stillness, involves the integration of both perceptual and physiological (sensory) signals. Our physiological and sensory systems send messages to the brain, which encodes information about our direction of travel, our orientation and the location of our body parts. Our perceptual system is based on our knowledge of the world and it is responsible for interpreting our physiological and sensory states.

Feeling dizzy

Feelings of vertigo can arise either due to conflicting, or ambiguous, physiological and sensory signals or due to a mismatch between physiological states and our perception of the world.

Have you ever lay on your side and rolled down a hillside, or been spun around by a group of friends, and then stopped moving suddenly? If so, you will probably have experienced the type of vertigo that comes from conflicting forms of sensory input. When your head spins around the endolymph moves in your vestibular canals sending "movement" messages to your brain and the brain then sends signals to the muscles of the eyes so that they move in the opposite direction to compensate. When we spin around quickly the endolymph moves to one side and then stops moving. When we finally come to rest and our head stops moving the endolymph whooshes back in the opposite direction before it settles down. As it moves back in the opposite direction the brain is still receiving movement signals and is sending compensatory signals to the eyes, activating the vestibular-ocular reflex. The stillness of our body in combination with the activation of the vestibular system and the VOR gives us the sensation that the world is spinning around us, and we have a sense of dizziness.

Imagine you're on a train journey. You have a window seat and outside you watch the world whizzing by. You have a sense of personal motion, such that you know you are moving past the scenery, which you know to be static. Now, have you ever had the experience where you've arrived at a train station and are sitting on a stationary train and there is another stationary train next to you. You're looking out of the window and you notice and feel yourself moving away. The train you are sitting on feels as if it is moving and this sense of movement is confirmed by what you see, which is that you are moving away from the stationary train on the other track. However, after a couple of seconds you notice that it is the train on the other track that has moved and your train is still waiting at the station. Yet you still had the sensation that you were moving. This form of vertigo is caused by your perception of the world. You see the world moving outside of the train window and based on your experience of sitting on moving trains and your knowledge of the world, you perceived that you were moving.

In these situations, where there are conflicting forms of sensory and perceptual input, the brain and the cognitive system need to resolve any ambiguity. It is not clear how this is done at a processing level. However, what is clear is that the way we resolve ambiguity caused by conflicting forms of sensory input may influence our perception of our feelings of vertigo.

Dancers' Dizziness Defying Techniques

Dancers learn to control feelings of vertigo throughout their training. Training gives dancers acute awareness of their proprioceptive system and it teaches them techniques that, when practiced, can help them to overcome feelings of disequilibrium, dizziness and vertigo. Extensive training might also lead to adaptation of some areas of the brain and such adaptations might also reduce dancers' sensitivity to vertigo.

Spotting

Every dancer who has been taught to pirouette will, I'm sure, have been taught a technique called spotting. A pirouette is, generally, a fast turn of the body. A single pirouette involves turning $360o$, that is, turning around and finishing facing in the same direction as when you started. A double pirouette involves turning $720o$, thus making two full turns of the body. Triple and quadruple pirouettes are commonplace and a professional ballet dancer performing Odile, the Black Swan, in Swan Lake can be expected to turn 32 fouette turns. A fouette turn is a special type of pirouette, sometimes called a whipped turn, because the dancer makes a whipping motion with their leg as they turn. In simple terms, a dancer stands on one leg and, as they turn their body, they move their other leg out in front until it is straight, then they move it to the side they then bend their leg so the toe comes towards the knee and then the movement starts again with the leg moving out to the front. This creates a whipping motion and is done while the dancer is turning, often en pointe.

Spotting during pirouettes involves the body and the head moving at different speeds. The body moves, approximately, at a constant speed whereas the head starts to move after the body has started to move. The head then whips around faster than the body and the dancer fixates their vision on a "spot" while the body catches up with the head. If a dancer is turning more than one pirouette then the head remains still for a fraction of a second while the body starts its next rotation and then the head overtakes the body again as it whips around. The technique is called spotting because a dancer must fixate on a spot, for example a mark on a studio mirror, on a wall or on a point in the auditorium.

The point of spotting is to help dancers maintain balance and to retain good spatial orientation. Having a visual fixation point as dancers turn helps to reduce nystagmus (small involuntary movements of the eye, such as those involved in the vestibular-ocular reflex) and it controls a dancer's perception of movement. The act of spotting gives dancers a moment of stillness in an otherwise continuous set of turns and it is thought that this provides an important context that can be used to overcome those feelings of vertigo that derive from messages being sent from the vestibular canals and the otolith organs.

When we watch dancers train or perform it is clear to see that they have mastered the negative effects of vertigo and dizziness and there is some experimental evidence to support this observation, which suggests that dancers experience less dizziness than non-dancers (Osterhammel et al., 1968; Nigmatullina et al. 2015).

Neural adaptation

Does training and experience lead to changes in the structure of a dancer's brains Yes, it seems that it does. The brain is made up of billions of brain cells called neurons. Each neuron has a cell body which receives input from numerous input branches (called dendrites) and the cell body then sends signals to other cell bodies or parts of the brain along an axon. Different areas of the brain are made up of either grey matter or white matter. The grey matter areas consist mainly of densely packed neuronal cell bodies and the white matter areas consist mainly of long-range axon tracts.

Hanggi et al. (2010) examined particular brain regions of 10 ballet dancers and 10 non-dancers and, using MRI brain scans, found significant differences in grey matter and white matter volumes. They found the ballet dancers had decreased volumes of grey and white matter in several areas of the brain that are associated with sensorimotor movement (for example, the premotor cortex, the supplementary motor area, the putamen and the corpus collosum). They also found that in the dancers there was a correlation between the age at which people started to dance and their grey and white matter volumes, suggesting that dancers are not born with different grey and white matter volumes in the brain but rather training in dance alters grey and white matter density in certain brain regions. Hanggi et al. conclude:

"...the structurally altered sensorimotor brain structures in ballet dancers, which were found in the present study, might represent the neural correlates of increased performance in organizing body movements into spatial patterns, controlling the body posture precisely, synchronizing their movements with regular and irregular rhythms, integrating proprioceptive information from several muscles and joints in order to generate a representation of the body in space, and in coordinating the body nearly perfectly, ranging from gross to very precise fine motor movements." (p. 1202).

Although Hanggi et al. didn't explicitly test for superior balance, or the perception or control of vertigo, in their sample of dancers their conclusions suggest that dance training/experience might mediate changes in those brain regions responsible for controlling body posture and proprioceptive information. Nevertheless, on the basis of the observation that dancers are able to control their body and balance, and considering the neurological changes observed by Hanggi et al. we should expect that dancers will have better balance than non-dancers and this should be observable in a range of situations.

Studies of Dancers and Balance

In simple tests of balance, it seems that years of training and neural adaptation has no positive effect on the balance abilities of dancers, and it may even make them more unsteady on their feet. This isn't what we were expecting.

In the simplest of balance tasks Perrin et al. (2002) asked ballet dancers, judoists and a control group to stand on a force plate with their eyes open and look at a point on a wall. While the participants stood still they measured the amount of body sway over a 20 second period with the expectation being that those with better balance should have the least amount of body sway. Perrin et al. then made the task more difficult. First, they added a sensory deprivation condition by asking participants to stand with their eyes closed, and then they added a "perturbation" condition such that they started to move the force plate while the participants tried to keep their balance. Using this technique they were able to assess the effects of visual, vestibular and proprioceptive contributions to balance.

In the eyes open condition with a stable platform Perrin et al. report that the balance of dancers was generally no better or worse than that of the judoists or the controls. A similar finding was also reported by Kuczynski et al. (2011), who found that, during a quiet-stance task, competitive dancers showed the same balance profile as non-dancers.

However, Perrin et al. report a different pattern of findings when they asked their participants to close their eyes. With eyes closed the balance of dancers was worse than that of the judoists and in one case worse than that of controls. This suggests that dancers make extra use of visual input in the control of their balance as compared with judoists and control participants. When they are relying only on proprioceptive input (as is the case when trying to maintain balance on a solid base with eyes closed) they are less able to keep their balance and consequently they experience greater body sway.

Kiefer et al. (2011) suggest that failure to observe superior balance in dancers may be due to lack of task difficulty. They argue that when a task is not demanding enough then it may mask superior balance performance. Using a one-legged balance task they show that dancers demonstrate more stable ankle-hip coordination than non-dancing controls during a visual tracking task. Stable ankle-hip coordination is thought to be indicative of enhanced balance because it shows precise, economical physical adaptation. Kiefer et al. conclude,

> "Dance training may reduce the number of constraints on ankle-hip coordination in order to enhance adaptability and flexibility of movement patterns. Dancer-like skill perhaps reflects an optimal level of deterministic coupling among movement system degrees of freedom." (p. 79).

This suggests that when you increase the difficulty of the task dancers rely on both visual and proprioceptive information for the maintenance of good balance.

In line with Kiefer et al.'s suggestion, Kuczynski et al. (2011) observed in a second condition of their study that dancers showed better balance than controls (in one aspect of balance) when a secondary cognitive task was added to a quiet-stance task (for example, the Stroop task, which involves looking at words printed in different coloured ink and naming the colour of the ink that a word is printed in. It's a difficult task because the words are themselves colour names and there is often a conflict between the name of the word and the colour of the ink in which it is printed. For example, if you saw the word PINK, you would have to name the colour of the ink in which it is printed, so the correct response would be BLACK). Adding a secondary task might be seen to make the primary task (balancing) more difficult. Kuczynski et al. (2011) found that dancers had better antero-posterior (forwards and backwards) balance than controls as they tried to name colour words during a balance test.

Kuczynski et al.'s (2011) finding may be related to the field dependent/independent theory of Witkins – see Crotts et al. 1996. Dancers are thought to be more field independent, suggesting that they see the visual environment as independent (from themselves) whereas non-dancers are more likely to be field dependent. Those people who are field dependent are likely to have poorer balance on the forward/backwards plane.

Further evidence for dancer's superior balance in challenging situations comes from Crotts et al. (1996) who compared the balance of dancers and non-dancers in a range of one-legged balance tasks that varied in difficulty. They found that although there was no observable difference in balance between dancers and non-dancers in the unchallenging tasks, for example, standing on one leg, eyes open on a flat firm and stable floor, there were differences in favour of dancers as the task became more difficult, for example standing on one leg, eyes closed on a foam step.

Crotts et al. suggest this is due to dancers being able to make use of additional somatosensory and vestibular information in the absence of visual information and pure vestibular information when both visual and somatosensory (proprioceptive) information is restricted (during eyes closed and standing on unsteady foam).

Spinning around

Research in the previous section has focused mainly on tasks that involve balancing while standing on one or two legs. A study carried out at Imperial College, London by Nigmatullina, Hellyer, Nachev, Sharp and Seemungal (2015) used a rather more dramatic method of measuring dancer's response to being off-balance. Nigmatullina et al. strapped dancers and non-dancers into a spinning chair and they recorded participant's eye movements and also how much participants thought they were moving as they went around in circles in the dark. The eye movement measure gave Nigmatullina et al. an indication of the vestibular-ocular reflex (VOR) and the subjective movement measure gave them an indication of participant's perceptual response to movement. Normally, perceptual responses are tightly coupled with reflexes and therefore allow researchers to predict one measure based on the other. In other words, a high degree of VOR is associated with a strong perception of movement and a low perception of movement is usually a good predictor of low levels of VOR.

Nigmatullina et al. recruited 29 dancers and 29 control participants and then spun them around in the chair. They observed the traditional relationship (correlation) between VOR and the perception of movement in the control participants, such that higher levels of VOR were associated with more perceived motion. However, in the dancer group the relationship between VOR and perception of movement was very different. They found that a dancer's perception of movement was not correlated with their VOR. In other words, a dancer's perception of movement is not driven by vestibular information alone and feelings of vertigo appear to be attenuated in dancers. Nigmatullina et al. conclude that "…an uncoupling of vestibular perception and reflex is an advantageous behavioural response to vestibular training…" (p. 6). In the case of dancers vestibular training has occurred in the dance studio with thousands of hours practicing balance and turns.

To locate the locus of the neuroanatomical differences between dancers and non-dancers Nigmatullina et al. scanned all of the participants using magnetic resonance imaging (MRI) and, like Hanggi et al. (2010), they measured grey matter and white matter volume. Compared to control participants ballet dancers showed reduced grey matter density in the vestibular cerebellum. In addition, there was a negative correlation between dance experience and grey matter density in the vestibular cerebellum, such that dancers with greater dance experience had less grey matter density. Nigmatullina et al. also observed some differences in white matter between dancers and controls. For example, in the control participants there was a positive correlation between perception of self-movement and cortical white matter microstructure whereas in the dancers there was no such relationship.

Summary

These findings suggest there may be two main effects of long term training in dance (ballet). Dance training appears to have an effect on the structure of the brain and also on dancers' perception of movement. It seems that these two effects work together to give dancers superior balance abilities More specifically it seems that ballet training reduces grey matter density in the vestibular cerebellum and this may support (or be a consequence of) the uncoupling of reflex and perceptual signals as they relate to lower feelings of vertigo in dancers. The second is that long term training in ballet may lead to resistance to feelings of vertigo because of changes to the white matter cortical networks associated with vestibular perception.

CHAPTER 10

Recreational Dance and Self-esteem

I attend a jazz dance class several times a week. I was waiting to go into class recently and I was speaking with three of the other dancers. They asked what I was working on and I told them about this section of the book. I told them about how little published academic evidence there was for a relationship between dancing and self-esteem. All three of them were quick to tell me that dancing definitely had an effect on their self-esteem. They gave me examples of how different classes made them either feel confident and fabulous or lumpy and worthless (so they mostly attend those that make them feel competent and confident), but overall, they told me, dancing made them feel good about themselves. It definitely had a positive impact on their self-esteem, and so it does with mine. Dancing can make me feel good about.

The Health Education Authority examined the link between participation in the arts and health. Their report, published by the Health Development Agency (2000), states, amongst other things, that engaging in arts-based activities improved participants' sense of wellbeing and self-esteem. The data they report seem compelling and conclusive. Across ninety projects 91% reported a development in people's self-esteem and 82% reported increased confidence. From these data, it would seem that participation in community arts-based projects (including dance activities) has a positive impact on people's self-esteem and confidence.

Anecdotal Evidence of improved Wellbeing

However, the widespread health benefits of engaging in arts-based activities reported by the Health Development Agency were based predominantly on anecdotal, non-empirical, evidence. Arts organisations were sent a questionnaire which, it was stated, explored the link between participation in the arts and health. Respondents (i.e. people who ran arts-based projects, rather than the individuals who took part in such projects) were instructed to complete a 31-page questionnaire as quickly as possible. The questionnaire comprised 12 sections with a mixture of closed questions, which required tick-box responses and some open questions, which allowed free text responses. One of the sections concerned relationship to health which respondents were told "maps your relationship to health and education (even if you've never analysed these links before)" (p. 36). The penultimate line of the instructions was "With your help we hope to put the link between arts and health firmly on the map." With such instructions, it is possible that respondents may have been susceptible to demand characteristics, such that they might give answers which are consistent with what the researchers are hoping to find.

Self-esteem and confidence data were captured in two questions. Self-esteem was assessed in Section D: Relationship to

Health and Wellbeing. Sub-section D1 asked "How does the project contribute to health improvements in the area?" One of the tick-box choices was "Develops people's self-esteem". Confidence was assessed in Section G: Monitoring and Evaluation. Sub-section G6 asked "We don't expect formal evaluation methods, but in your opinion, do results suggest any of the following?". One of the tick-box options under Personal Development was "Increased Confidence".

It was using these methods that 91% of project managers ticked the develops people's self-esteem box and 82% ticked the increased confidence box. These data should not be taken to imply that 91 and 82 percent of project participants realized a positive change in their self-esteem and confidence, respectively, as a consequence of engaging in arts-based activities.

Nevertheless, what these data do tell us is that the majority of people who ran arts-based projects, and who completed the questionnaire, feel that engaging in these projects has a positive effect on participant's psychological wellbeing and self-esteem. The next step is to ask whether these feelings and informal observations are confirmed in more empirically based studies.

Empirical Studies of the Benefits of Recreational Dance to Self-esteem.

There are several published studies that report a relationship between engagement with recreational dance and psychological wellbeing, including self-esteem (Beaulac, Olavarria & Kristjansson, 2010; Blackman, Hunter, Hilyer & Harrison, 1988; Burgess, Grogan & Burwitz, 2006; Gardner, Komesaroff & Fensham, 2008; Quin, Frazer & Redding, 2007; South, 2006). Unfortunately, this collection of research does not unanimously support the conclusions drawn from the Health Development Agency's (2000) report. It does, however, raise some interesting questions about the relationship between participation in recreational dance and wellbeing.

The aerobic dance versus swimming study

Burgess, Grogan & Burwitz (2006) provide the strongest empirical evidence to date that engaging with recreational dance has a positive impact on aspects of self-esteem. However, their findings suggest that the effects may be limited and their conclusions should be qualified. Burgess et al. set up an experiment to examine the effects of 6 weeks of exercise (either aerobic dance or swimming) on body image dissatisfaction and physical self-perceptions, including global self-esteem, of previously non-exercising girls. Fifty 13-14 year olds, with a profile of high body-image dissatisfaction, low physical self-perceptions and low physical activity levels were measured using the Body Attitudes Questionnaire (BAQ; Ben-Tovim & Walker, 1991), The Children and Youth Physical Self-Perception Profile, which includes a measure of global self-esteem (CY-PSPP; Whitehead, 1995), and the Leisure Time Physical Activity Questionnaire (LTPAQ). The participants were measured at three time points: Pre-test (before the intervention), Mid-test (at the mid-point of the intervention) and Post-test (after the intervention).

All participants took part in a twelve-week exercise program; comprised of six weeks of twice weekly dance aerobics classes and six weeks of twice weekly swimming sessions. All sessions lasted 50 minutes. Before the start of the study participants were randomly assigned to one of two groups: A dance-first group and a swim-first group, such that the dance-first group did six weeks of dancing following by six weeks of swimming and the swim-first group started with six weeks of swimming. This is called a cross-over design and is used so that participants act as their own controls. There were no differences in pre-test scores on any of the measurements between the dance-first and swim-firm group.

For the sake of addressing the question of whether dancing can increase a person's self-esteem we are interested in the following patterns of results.

If exercise has no effect on a person's self-esteem then we should not expect to see any changes in the self-esteem of participants over time.

If exercise has a positive effect on a person's self-esteem then we should expect to see positive changes in the self-esteem of participants in both groups over time.

If one form of exercise (e.g. dance) has a greater effect on self-esteem than another (e.g. swimming) then we should expect to see positive change in self-esteem following six weeks of one form of exercise but a smaller change (or no change) in self-esteem following six weeks of the other.

Statistical analyses were carried out on the effects for body attitudes and physical self-perceptions. Effects relevant to aspects of self-esteem are reported here and the associated means and standard deviations are shown in Table 10.1

| | Time of Measurement | | |
	Pre	Mid	Post
Attractiveness			
Dance-first[a,b]	14.0 (2.51)	18.8 (3.56)	14.5 (3.35)
Swim-first	14.2 (3.07)	14.1 (2.81)	16.8 (3.51)
Feeling Fat			
Dance-first[a,b]	36.0 (8.71)	30.7 (8.87)	36.6 (9.00)
Swim-first[a,b]	36.2 (11.02)	41.6 (10.29)	34.9 (6.98)
Body Attractiveness			
Dance-first[a,b]	2.24 (.41)	3.19 (.27)	2.34 (.54)
Swim-first[b]	1.96 (.40)	1.92 (.50)	2.39 (.31)
Physical Self-worth			
Dance-first[a,b]	2.38 (.56)	3.08 (.31)	2.63 (.60)
Swim-first	2.32 (.52)	2.29 (.62)	2.51 (.40)

a = significant difference between pre and mid.
b = significant difference between mid and post.

Table 10.1: Showing selected data from Burgess, Grogan & Burwitz (2006) on Body Attitudes (attractiveness and feeling fat) and Physical Self-perceptions (body attractiveness and physical self-worth) as a function of time of measurement for each group.

Attitudes and perceptions of attractiveness were measured in both the BAQ and the CY-PSPP, as attractiveness and body attractiveness respectively. Similar patterns of results were observed for both measures, such that significant group by time interactions were observed for attractiveness ($F_{2,96} = 18.24$, $p < .0001$) and body attractiveness ($F_{2,96} = 45.36$, $p < .0001$).

For the BAQ attractiveness the pattern of data was as follows: for those in the dance-first group there was a significant increase in attractiveness from pre- to mid-testing. That is, scores for attractiveness were higher after six weeks of dancing. Attractiveness scores returned to pre-dancing levels after six weeks of swimming. For those in the swim-first group there were no significant changes in attractiveness over time. This means there were no changes after six weeks of swimming or after six weeks of dancing.

It is not clear why the dance-first group improved after six weeks of dancing but the swim-first group did not. If the increase in attractiveness in the dance-first group was something to do with the dancing, per se, then we should have expected to see changes in attractiveness in the swim-first group after they had danced for six weeks.

For CY-PSPP body attractiveness the pattern of data was as follows: there was a significant increase in body attractiveness for both groups after six weeks of dancing. In the dance-first group there was a significant reduction in bodily attractiveness after six weeks of swimming (marked by a return to pre-testing levels) and for the swim-first group there was no change in body attractiveness after six weeks of swimming. This finding appears to show that self-perceptions of body attractiveness are significantly improved after six weeks of dance participation.

Attitudes about feeling fat were measured as part of the BAQ. There was a significant group by time interaction for feeling fat ($F_{2,96} = 13.74$, $p < .0001$), such that there was a reduction in feeling fat scores following six weeks of dancing and an increase in feeling fat scores following six weeks of swimming. For the dance-first group feeling fat scores significantly decreased after six weeks of dance and then returned to pre-intervention levels after six weeks of swimming. For the swim-first group feeling fat scores increased after six weeks of swimming and then returned, approximately, to pre-intervention levels after six weeks of dancing. This suggests, for the swim-first group, no improvement from initial baseline measures on feeling fat following six weeks of dancing.

Self-perceptions of physical self-worth were measured as part of the CY-PSPP. There was a significant group by time interaction for physical self-worth ($F_{2,96} = 10.14$, $p < .0001$), such that for those in the dance-first group there was a significant increase in perceptions of physical self-worth from pre- to mid-testing. That is, scores were higher after six weeks of dancing. Physical self-worth scores returned to pre-dancing levels after six weeks of swimming. For those in the swim-first group there were no significant changes in physical self-worth over time. This means there were no changes after six weeks of swimming or after six weeks of dancing. Again, it is not clear why the dance-first group improved after six weeks of dancing but the swim-first group did not.

Burgess et al. (2006) conclude that their study presents a strong case for the positive psychological benefits of engagement in dance for 13-14 year old girls, who initially have a poor self-image.

There were differences in the patterns of data following six weeks of dancing and six weeks of swimming, such that dancing seems to be associated with greater improvements in different aspects of self-esteem. Burgess et al. suggest this may be due to the dance environment being less competitive and less threatening, and therefore more supportive, than the environment surrounding swimming training. This, they argue, may provide the appropriate environment and opportunities for young women, especially those with a poor self-image, to feel better about themselves.

One of the strengths of this study is in its design. Using a cross-over design, where participants acted as their own control and where all participants took part in both interventions in a different order, this study was able to show that the dance intervention led to changes in self-worth. It also showed that dance-induced changes in self-worth are temporary, in as much as they are brought on by six weeks of dancing.

This study is far from conclusive though and it raises issues for future research. For example, the participants in this study had high levels of body-image dissatisfaction, low physical self-perceptions and low physical activity levels. Six weeks of dance for teenage girls with this profile seems to have a positive impact. However, we don't know what benefit will be achieved from starting to dance for people who have a different pre-intervention profile, such as girls (or boys) with even lower body-image dissatisfaction or higher physical self-perceptions. It is possible that for dance to enhance aspects of self-esteem the person who dances must have low self-esteem to start with, or they must have low physical activity levels.

This study also raises the question of whether it is not just the dancing that has led to the improvement in self-perception but rather something about the characteristics of the dance and the way it was taught. The authors suggest that some forms of physical activity are better than others at developing feelings of competence and confidence in adolescent girls. There are, of course, many different types of dancing and motivations for dancing (or teaching dance). It appears to be the case that the aerobic dance used in the current study fostered feelings of competence and cooperation amongst the participants. It is possible that different outcomes might have been observed had the dancing challenged the competence levels of the girls, for example, if it had been technically challenging or had a large memory component, or if it required precision or a certain degree of performance quality.

Burgess et al. incorporated a design which meant that participants acted as their own control, however, they acknowledge that the presence of a true control group, that is a group of people who did not take part in any physical activity, would have helped them to pinpoint the benefits of dance more precisely. In addition, they point out that six weeks of aerobic exercise may not be sufficient to lead to sustainable differences in body attitudes and physical self-perceptions. A longer-term study is therefore recommended. Blackman, Hunter, Hilyer & Harrison (1988) had carried out a study which addressed these methodological issues, such that they measured a passive control group and took measures over a slightly longer time frame than Burgess et al. Unfortunately, the study by Blackman et al. raises more questions than it answers concerning the relationship between engagement with dance and self-esteem.

The high school dance team study

Blackman, Hunter, Hilyer and Harrison (1988) carried out a study to examine the extent to which female physical fitness and self-concept are affected by participating in a high school dance team. This paper is useful to the question of whether engaging in recreational dance has a positive impact on self-esteem because Blackman et al. directly assess changes in different aspects of self-esteem in a sample of 14-18-year-old girls before and several months after dance team participation. They also tested a control group who did not dance but, and it's not clear why, they only tested the control group once.

They used three specific instruments: The Coopersmith Self-esteem Inventory, the Tennessee Self-concept Scale and the Body Cathexis Scale.

The Coopersmith Self-esteem Inventory (Coopersmith, 1968) is a 50-item measure of attitudes about oneself that was originally designed to measure the self-esteem of children. It includes a list of 58 statements, some of which have a negative valence or maning, such as, "I find it very hard to talk in front of a group" and some have a positive valence or meaning, such as "I am proud of my work". Respondents are asked to state whether each statement is "like me" or "unlike me". Self-esteem is calculated by giving points to positive valence items that were answered with "like me" and negative valence items answered, "unlike me". A higher score indicates higher self-esteem.

The Tennessee Self-concept Scale is a 100-item measure of self-concept. Items were given as statements, such as, "I am too sensitive about the things people in my family say" and "I am a nobody" and respondents rate each statement on a 5-point scale, ranging from "Always False" through a mid-point rating of "Partly False and Partly True" to "Always True". The 100 items load on to six factors: physical self, moral-ethical self, personal self, family self, social self, and self-criticism. A separate score is given for each factor, plus a total score.

The Body Cathexis Scale is a domain specific measure of subjective wellbeing. Cathexis refers to the amount of emotional energy that is applied to a particular object. In the case of this instrument cathexis refers to the amount of emotional energy people apply to their bodily parts and processes. It is a domain specific measure because it is used to assess the extent to which people are satisfied with various parts or processes of their body. Respondents rate each body part/process on the following scale:

1 = Have strong feelings and wish change could somehow be made
2 = Don't like but can put up with

3 = Have no particular feelings one way or the other
4 = Am satisfied
5 = Consider myself fortunate.

Blackman et al. tested those in the experimental group across a five-month interval (June and then October). During this time, those in the experimental group practiced with the dance team 10-15 hours per week and attended a 40-hour summer intensive dance camp. From September, the dance team practiced for 1 hour each day after school. As you can see the participants in this study did considerably more dance than the participants in Burgess et al (2006), who danced for just 100 minutes per week. The control group did not take part in any organized extra-curricular activities. Blackman et al. used a slightly strange experimental design, such that their experimental group were tested twice, before and after forming a dance team, and their control group was tested just once, at the same time as the dance team was tested for the second time.

Effects relevant to aspects of self-esteem are reported here and the associated means and standard deviations are shown in Table 10.2. Blackman et al. found that the dance team members showed significant improvements in two subscales of the Tennessee Self-concept Scale (physical self and social self).

There were no other differences when comparing pre- and post- intervention measures for either The Coopersmith Self-esteem Inventory or the Body Cathexis Scale. In addition, there were no differences in any aspect of self-esteem between members of the dance group (post dance team participation) and the control group ($p > .05$ in all cases).

	Time	
	Pre	Post
Physical Self		
Dance-team[a]	67.5 (7.1)	72.0 (9.1)
Control		70.9 (8.0)
Social Self		
Dance-team[a]	67.0 (6.6)	71.1 (7.3)
Control		68.0 (10.3)

a = significant difference between pre- and post-.

Table 10.2: Showing selected data from Blackman, Hunter, Hilyer & Harrison (1988) on the Tennessee Self-concept Scale (Physical Self and Social Self) as a function of time of measurement for each group.

There were no other differences when comparing pre- and post-intervention measures for either The Coopersmith Self-esteem Inventory or the Body Cathexis Scale. In addition, there were no differences in any aspect of self-esteem between members of the dance group (post dance team participation) and the control group (p > .05 in all cases).

So, what conclusions can we draw from this study? If we take the profile of findings at face value then we might conclude that engaging as an active member of a dance team positively changes a person's physical self-concept, that is, how they view their health, appearance and physical skill, and it also positively changes a person's social self-concept, that is, how they see them self in relation to their peers. These are potentially important changes, particularly for adolescents who are dealing with changes to their physical identity and who are likely to be interacting with changing peer groups. These findings also suggest that there are some longer term and persistent positive impacts on self-concept of participating in a dance team.

However, we must treat these findings with caution, with small sample sizes, unsophisticated statistical and lack of a consistent control group all indicate that this study needs to be repeated to provide a clearer answer to the hypothesis.

My first concern is with the experimental method and the very small sample size. In the experimental group, there were just 8 girls. A very small sample size like this leads to two problems. First, as Blackman et al. acknowledge, it prevents the use of sophisticated statistical analysis. This is a problem for the current research because the researchers carried out 60 comparisons of pairs of means using a t test (with, as far as I can tell, no Bonferroni correction for multiple comparisons). This increases the chance of making a Type I error (that is falsely rejecting the null hypothesis). As a consequence, we are left with a question mark hanging over the findings of this research because we cannot be certain that the positive changes observed in self-concept, for example, didn't happen by chance. The use of the statistical procedure MANOVA may help the authors overcome this uncertainty.

The second problem with this small sample size is in the researcher's failure to observe changes in self-esteem and body cathexis (and other aspects of the self-concept). We cannot conclude with certainty that participating in a dance team does not change a person's self-esteem because the small sample size may have meant there was insufficient power in the process to show an effect, even if one existed. This is a Type II error (that is falsely accepting the null hypothesis). In this study one of the hypotheses was that engaging in a dance team would increase a person's self-esteem (thus the one-tailed t test). Therefore, the corresponding null hypothesis would be that engaging in a dance team does not increase a person's self-esteem. Do we really think that testing just 8 people is sufficient to conclusively accept this null hypothesis? I think we need a larger sample and a better design before we can answer this question.

It is not clear to me why Blackman et al. didn't test their control group over the same time period as the experimental group. They found no difference in any of the psychological measures between the experimental and the control group at post-intervention testing, not even in those areas where the experimental group improved. In addition, the lack of a control group at all time points, or multiple control groups, doesn't allow us to determine the locus of the observed changes in self-concept. We might tentatively conclude that these changes were due, at least in part, to dance team participation. However, without testing proper control groups we do not know whether the real effect was due to factors such as the social bonding that came about because the girls went on a summer camp together, or due to the attention they got from intensive coaching over an extended period or because of the girls' increased physical fitness (this last point was acknowledged by Blackman et al.).

Blackman et al. measured the girls at two time points. Time 1: in June, prior to any dance team participation and Time 2: in October (of the same year) after dance team participation. It is disappointing that Blackman et al. didn't take more time-based measures of the participants as they progressed through the programme, especially as the level and type of activity changed across the five-month period. It would have been informative to the relationship between dance and self-esteem to understand how self-esteem changes as a function of different types and intensities of dancing. Try outs for the dance team were in May and once selected the first sets of measurements were taken in

June. The dance team then practiced for 2-3 hours 4-5 times per week for the summer months. This is a lot of dancing and is likely to be demanding in terms of energy, concentration and motivation. I would like to have seen how self-esteem changed across this sustained period. The participants then went on a four-day summer camp where they danced for 10 hours per day. Again, this is an intensive period of dance and cohabitation that is likely to have impacted on how the girls felt about themselves and each other. Finally, once back at school in September the girls practiced for one hour every day after school and then performed dance routines in public at football games and at pep rallies (a meeting, typically before a sporting event, to encourage enthusiasm).

It should be noted that some of the improvements in self-concept observed in the dance & swimming study by Burgess et al. (2006) were attributed to the participants dancing in a non-competitive, supportive environment, where they were able to feel competent and confident. There will have been times in the dance team participation study by Blackman et al. (1988) where the girls might also have felt they were in a supportive, safe environment where they could feel competent and confident and it would have been useful for Blackman et al. to have taken additional measures at this time. Simply measuring at the beginning and at the end of the process, when the girls would have been in a public, competitive situation doesn't give us a clear picture of the potential changes that might be brought about by the individual components of preparing for dance team participation.

This is an interesting study, but due to its methodological weaknesses, and tiny sample size, it doesn't tell us as much as we need to know about the relationship between dancing and self-esteem.

Creative dance in a non-competitive environment study

There is a third study, by Quin, Frazer and Redding (2007), which tests a much larger sample size (n = 348) than both Burgess et al. (2006) and Blackman et al (1988), and used 10 weeks of creative dance, in a non-competitive environment. Quin et al. measured psychological wellbeing, quantitatively and qualitatively, before and after the intervention. If dance has a positive impact on self-esteem then a project of this size and type should clearly show it.

Quin, Frazer and Redding (2007) examined the health benefits of engaging with creative dance. They examined both physical and psychological wellbeing. I will focus on their findings and conclusions as they relate to psychological wellbeing.

Quin et al.'s main analysis used a repeated measures design to examine changes in the psychological wellbeing of a group of 348 11-14 year olds, before and after taking part in a 10-week programme of creative dance. Psychological wellbeing was assessed using unspecified measures of self-esteem, intrinsic motivation and attitudes towards dance. The paper gives no details of the mean scores on these measures for before and after the intervention. However, they do report that there were no statistically significant differences between the pre- and post-intervention scores on any measure of psychological wellbeing. Therefore, the main conclusion that can be drawn from this research study is that engagement with a 10-week programme of creative dance did not improve the self-esteem, or other aspects of psychological wellbeing, of the participants. However, this is not how Quin et al. report their findings.

Quin et al. report that "Psychological wellbeing results found positive adaptations in all areas assessed..." (p. 32) they go on to conclude that: "Psychologically the positive responses in the well-being assessments suggest that creative dance has the potential to elicit positive effects on self-esteem, motivation and attitudes

175

towards dance." (p. 32-33). The key word in this sentence seems to be "potential". There is evidence that the majority of the female participants found the dance programme fun, exciting and enjoyable but I don't think this, taken with a set of non-significant numeric trends, is strong enough evidence to claim that creative dance improves the psychological wellbeing of young people in general. I think the picture from this research is that creative dance has the potential to improve the self-esteem of some people, perhaps people (females?) who enjoy engaging with dance. It is clear, as we saw in the research into dance and cardiovascular functioning, that more studies are needed in order to understand the complexities of when dance can improve self-esteem for different groups of people.

The premise of Quin et al. is basically sound. They used a large sample of young people and measured their psychological wellbeing (including their self-esteem) before and after engaging in 10 weeks of creative dance. There was no statistically significant change in self-esteem reported over the 10-week period. The majority of the female participants enjoyed it. The claimed, yet un-described, numerical trend is intriguing. Was the p value for self-esteem close to .05? What were the confidence intervals? Was there an interaction between changes in self-esteem and gender, age or degree of enjoyment? It might be the case that enjoyment of creative dance mediates changes in self-esteem. Alternatively, no significant changes in self-esteem might have been the result of ceiling effects in this sample. Perhaps they were a sample of happy, confident, well-adjusted young people for whom no or little increases in self-esteem were likely or necessary.

So, across three studies (Burgess et al., 2007; Blackman et al., 1988; Quin et al., 2007) that used quantitative research methods to examine the link between self-esteem and participation in different forms of recreational dance (dance aerobics, dance team and creative dance) there is only one set of data which shows how participation in dance can temporarily increase the self-esteem of low self-esteem girls. Taken together, these studies do not present a clear picture of how dance can improve self-esteem and they most certainly do not confirm the findings of the Health Development Agency (2000) report, which states that engaging in arts-based activities improved participants' sense of wellbeing, self-esteem and confidence.

Although the quantitative evidence is suggestive, yet inconclusive of a link between recreational dance and self-esteem, there is a body of qualitative evidence which appears to show that engaging with recreational dance can increase feelings of self-esteem and confidence in different groups of people. I will now describe the findings of South (2006) and Gardner, Komesaroff and Fensham (2008) in the hope that their qualitative methods can shed additional light on the issue.

Qualitative Studies of the Health Benefits of Recreational Dance

Community arts projects study

South (2006) reports a qualitative evaluation of the Bradford Arts for Health program, a community arts health program, which included aspects of dance and other arts activities. The Bradford Arts for Health program primarily involved the delivery of three community projects. These were a music and movement project for mothers and babies (aged between 0-4 years), a drama workshop exploring arranged marriages for Asian women aged between 14 and 17, and a Community Centre project for 8 to 13 year olds, which involved workshops on graffiti art, DJ skills, urban dance and mosaic/artworks running through the summer holidays. South (2006) reports that 14 people took part in the Drama Workshops (arranged marriages) but does not report the number of participants in the other projects.

Individual evaluation processes were designed for each project and these were based on each project's goals, objectives and indications of success. The declared aim of the evaluation was to get a detailed understanding of how and why the projects worked and, based on interviews with 16 people, the intention was to capture the experience and views of the people involved in the projects. However, it should be noted that the young people who took part in these projects did not contribute to the evaluation. Therefore, the evaluations are based on project workers, artists and other people involved in the implementation and delivery of the projects.

Semi-structured interviews were carried out which covered three main areas: 1. The development and implementation of activities, 2. Perceptions of the impact of activities, and 3. The contribution of the programme. Thematic analysis was used to identify cross-cutting issues. As such, and unlike the quantitative studies discussed above, there was no explicit, systematic measurement of self-esteem.

Broadly speaking there was a positive uptake and response to the Drama Workshops and the Community Centre projects. The response to the mother and toddler music and movement workshops was less positive. The Community Centre project involved the greatest amount of dance (urban dance). Based on the thematic analysis of the 16 interviews South (2006) reported that the young people in the Community Centre project had increased their self-esteem and confidence, and people in the Drama Workshops were able to identify issues affecting their mental and sexual health. Again, it should be noted that not all of the 16 interviewees were involved with both, or perhaps either, of these projects. Nevertheless, South reports that one of the interviewees spoke of a potential boost to self-esteem of people involved in the Community Centre project. In other words, one of the interviewees thought that engagement in this type of project will have a positive impact on a young person's self-esteem.

Unfortunately, this evidence is not strong enough for us to understand whether engaging in a partially dance-based recreational activity improves a person's self-esteem. Like those respondents to the Health Development Agency (2000) report, people working in arts-based projects assume a link between the arts in general, and dance in particular, and psychological wellbeing and expect to see such a link either prospectively or retrospectively. Nevertheless, the evidence is not conclusive. There are links to other areas of psychology here. Consider the work of Bartlett (1932) who found that memory and interpretation of stories was based on what people expected to see. Perhaps it is the case that people expect to see changes in the self-esteem of others following engagement with arts based (and dance) activities.

The limitations of the study are clearly articulated by South (2006). She states that although using a qualitative approach allowed issues to be explored in depth there were several limitations. First, she writes that the evaluation does not provide evidence that can be generalized. It is clear, I think, that the evaluation is extremely limited. It is based on 16 interviews and it is not clear how many of the interviewees spoke about each of the three projects. The second, related, limitation South discusses concerns the make-up of the interview panel. Artists were involved in some of the data collection and this, she suggests, could have led to biases in the interpretation of the responses.

There are, potentially, financial consequences for projects that are evaluated negatively, such that projects may lose their principal funding sources if they are not seen to be delivering on their specified objectives. Therefore, the use of artists, whose jobs may depend on receiving positive feedback, as interviewers and data handlers may lead to distorted findings. I have already mentioned South's third limitation, that is in the failure to ask for feedback, or measure changes, in the participants themselves. South argues this was because some of the participants were potentially vulnerable, although she does not say in what way, and it was therefore not appropriate to engage them in extensive evaluation. I am a little surprised by this. Psychologists have many ways and established practices for interviewing vulnerable (young) people that are sensitive and safe. Finally, it was noted that this study does not include any measurement of changes in health over the course of the programmes.

Despite the declared limitations of this project South (2006) argues that "The potential benefits of arts for health is undisputed" (p. 166). I'm not so sure about that. If we take such benefits of arts for health for granted then we are at risk of basing our assumptions on faith rather than evidence.

Recreational dance study including tango, Greek, ballet, hip-hop, ballroom, belly and flamenco

The final paper to be discussed in this section describes a project that has examined young people and adults' engagement with a range of dance styles and, again, has taken a qualitative approach. Gardner, Komesaroff and Fensham (2008) carried out semi-structured in-depth interviews of ten people between the ages 14 and 26 who attend dance classes outside of an institutional setting. The interviews attempted to identify the motivations for attending dance classes, participant's experiences of attending such classes and the personal perceived impact of attending these classes on body image, health and wellbeing. The participants, both men and women, attended established recreational dance classes in a range of dance styles including: Tango, Greek, ballet, hip-hop, ballroom, belly and flamenco.

Thematic analysis of the transcribed interviews revealed five themes. Quoting directly from Gardner et al. (2008) these were:

> 1. Dance classes foster respect for physical activity and expertise acquired over the long term.
> 2. Dance class participants gain self-confidence with respect both to their bodily experiences and social-relationships.
> 3. Dance classes increase respect between older and younger people in a physical activity context.
> 4. Dance classes are sites for exploring or maintaining social, community/cultural, recreational or intersubjective values.
> 5. Dance class participation can involve experiences of self, body and world that lie beyond the everyday.

With respect to self-esteem it is clear that the second theme, dance class participants gain self-confidence with respect both to their bodily experiences and social-relationships, is the most relevant to our current discussion. Gardner et al. report that all ten of the participants felt that dancing made them feel good about themselves, such that it lifted their spirits and energies. The participants also commented that dancing was good for their posture and self-esteem. Based on the design of this study and the methods used for analysis we don't know how many of the participants made this comment, or what type of self-esteem they were talking about. As such we don't know whether people felt that dancing changed something about their self-worth at a fundamental, global or trait level, or whether the act of dancing gave them a temporary boost in state self-esteem that was related, for instance, to their dance-related competence.

Several participants apparently commented that dancing made them more comfortable with other people and with intimacy. We might assume these comments were made by people who attended partnered dance classes, such as tango and ballroom, in which case the benefit derives directly from a central aspect of the particular dance style. However, it might also be the case that that these feelings derived from non-partnered dance, in which case it would show that dance has effects beyond the characteristics of particular dance styles. Without the detail we are not able to draw firm conclusions.

Dancing is not just a physical activity, it is also a social, emotive and cognitive activity. The cognitive, thinking, aspect of dance may also contribute to changing a person's self-esteem. Gardner et al. quote a 17-year-old female who speaks about the cognitive and emotional aspect of dance as follows:

"Different dances have different moods so, like for example the Rumba you have to sort of pretend that you are in love, it's a whole story that you create on the floor, it's not just the dancing you do and like we have a lot of things that I guess where you have to imagine, you've got to work well with your partner and to be able to read them."

Based on this type of comment we might expect the changes people report in terms of their self-esteem to relate to social aspects of self-esteem. It would be interesting and informative to use

Despite the small sample size, wide age range and lack of specificity concerning the relationship between dance and psychological wellbeing Gardner et al. (2008) start their discussion with the bold phrase that their study "has shown that participation in community dance classes can enhance respect for physical activity and improve confidence…" Once again, this is an over statement. They have not shown that participation in community dance classes improves anything, and they cannot draw this conclusion because they did not measure people at time one and then again at time two and measure the change. Neither did they use a control group of, say, non-dancing physically active people and show that such people had lower psychological wellbeing than their dance-class attendees.

Gardner et al. do point out two weaknesses with their study. First, they concede that their sample size was too small to be representative and say that they may not have reached thematic saturation. This means that had they interviewed more people other themes may have emerged. They also make the very important point that the five themes that emerged from the interviews may have represented the nature and characteristics of the participants per se rather than the nature and characteristics of people who dance in general. As such, they are not able to conclude that these themes emerged as a consequence of engaging in recreational dance.

Summary

In this chapter six published reports have been examined as they relate to the question of whether engaging with dance-base activities can lead to a positive change in self-esteem. Although several of the studies make grand claims regarding this relationship, for example stating that dance has a positive impact on a person's psychological wellbeing, the evidence for such claims is, at best, inconclusive. There is very little methodological consistency across the studies. They each use, as far as we can tell, different tools to measure self-esteem and the samples are drawn and tested differently. It is interesting that changes in self-esteem are reported in those studies using semi-structured interviews, two of which were based on third party assumptions (e.g. Health Development Agency, 2000; South, 2006) and yet no changes are observed when specific measures of self-esteem are used (such as in the case of Blackman et al., 1988 using the Coopersmith Self-esteem scale.). It is clear that many of the studies had very small sample sizes and this may account, in part, for the variability in findings across the studies. However, a failure to detect significant and meaningful changes in self-esteem was reported in the study with the largest sample (Quin et al., 2007) so perhaps the small sample size is missing nothing. Perhaps dancing just has no effect on self-esteem. If it wasn't for Burgess et al. (2006) we might draw this conclusion, but their findings seem to hint that something is going on. What was interesting about their study was that they observed changes in self-esteem in a sample of young women who started the project with low self-esteem. It was also noteworthy that the positive changes appeared to be short lived. Burgess et al. compared dancing with swimming and they suggest that the dance sessions were less competitive, less threatening and more supportive than the swimming sessions. Therefore, the observed changes in self-esteem might be due to the presence of these conditions in a learning environment rather than having anything to do with the act of dancing per se. The findings of Burgess et al. also suggest that changes in self-esteem are not just dependent on the activity that is currently being engaged with but also by the profile of activities. Gardner et al. report that all of their (ten) participants said that dancing made them feel good, and for some it is thought to have raised their self-esteem.

Perhaps the issue here, taking in to account everything in this chapter, is one of anecdote versus evidence. It is clear that people, anecdotally, feel that dance either raises their own or other people's self-esteem. However, the experimental evidence does not support this conclusively. Perhaps we need better scientific methods for measuring changes in self-esteem.

CHAPTER 11

The Communication of Emotion

The scientific study of dance and the creation and performance of artistic dance are worlds apart. While the former is constrained by what has gone before, minimal and theory-driven the latter is free, expansive and driven by creative ideas and sometimes by a desire to break away from everything that has gone before. By combining these two ways of looking at dance we can learn something about how we use dance as a language and as a form of communication. In this chapter I discuss the scientific literature on how we recognise emotions through dance and body movement. I introduce some lab-based experimental and cognitive neuropsychological research and I then discuss the issue of how choreographers break away from the constraints of science and describe the performance of 23 Feelings in Dance.

War Dances

82,000 people gather to watch fifteen men dance. The men are tough. Some are short and broad with broken noses and others are lithe and light-footed. This is a major competition and there are no female dancers here. The men are about to perform a dance called the Haka. The dancers are the New Zealand rugby union team and no, this is not a celebrity dance competition. They don't perform the Haka on Strictly Come Dancing.

The New Zealand team dance the Haka before every international rugby match and it is, in essence, a war dance. There are several varieties of the Haka. There's the "peruperu", a choreographed dance performed before battle, to evoke the god of war, and there's the "ngeri", an un-choreographed dance which motivates people psychologically. So, the Haka can speak to the gods and the individual. The Haka involves facial contortions, sticking out your tongue, making your eyes bulge, stamping your feet, and slapping your body and it can be accompanied by cries, grunts and chanted words. The Haka gives the rugby team energy and it makes them feel powerful. The dance also instils fear in the opposition, just as it would have done when it was danced before battles to the death. The function of the original haka peruperu, or dance of war, was to give opponents one last chance to back out, before battle to the death began. War dances that were well organised and executed in perfect time by all the dancers (warriors) were considered lucky. These really are ancient rituals. It is thought that the dances were passed down from one generation to the next and they may have originated as early as the 13th century, when the first Polynesian settlers arrived in New Zealand.

In some forms of the Haka the dance unifies the dancers, it puts them in a single frame of mind, and as such the Haka can be seen as a vehicle for temporarily changing the way people think. Dancing the Haka, therefore, helps people to communicate with the gods, terrify observers and it can make those who perform it feel united, powerful and invincible.

If the New Zealand rugby team are trying to communicate something to their opponents by dancing the Haka before a match then it is clear that they can only effectively do this if it is possible to communicate emotions through dance and body movement.

Communication through body movement is a two-way process. First, we must be able to transmit a message through our body movement and second, someone else needs to be able to receive, interpret and understand that message. When we speak with our voice we must know how to speak in a particular language to transmit a message and the listener needs to be able to hear the sounds of words to receive the message, and convert the sound of your voice into words to interpret the message and know the meaning of the words to understand the message.

With both written and spoken language we rely on knowing and understanding the structure of formal languages (such as French, English, Arabic, and Cantonese). When we communicate using our body we can either use a formal structured dance "language", like classical ballet or an Indian classical dance form such as Bharatanatyam, where certain iconic movements and gestures are used to communicate particular ideas or feelings, or we can use informal, natural or abstract movements. When, in dance, we use a formal structured dance to communicate a feeling or mood then both the person transmitting the message and the person receiving the message must have knowledge of the language to be able to communicate effectively. However, when we use natural or abstract movements there is no necessity to explicitly "know" the language in advance.

Psychologists are interested in a wide range of questions concerning dance and the communication of emotion. These include questions such as, what emotions can be communicated through dance, what information about the body and the way it moves is required to communicate such emotions, what are the psychological, biological, and/or neurological mechanisms which support the recognition of emotion? I'm interested in all of these questions but as a Dance Psychologist I've looked at these questions from two different points of view. The first is from the perspective of a scientist, whereby I want to understand how we communicate emotion and feelings through dance when everything is carefully controlled in a laboratory. The second is from the perspective of a dance artist, whereby I want to understand how makers of dance create movement which portray emotion to a particular audience. I'll give you the science bit first, then the dance as art-performance bit.

Communicating Emotion – a Scientist's Approach

When we see a picture of a happy/smiley face we generally know what it means. The same is true if we see a picture of a grumpy face. We can tell the emotion the person in the picture is feeling, or trying to communicate, automatically. Charles Darwin (1872) suggested there are universal facial expressions associated with different emotions, meaning that people from different parts of the world use the same facial expressions to communicate emotions, such as happiness, sadness and disgust. Therefore, signalling emotions with facial expressions is an innate activity. In other words, it suggests that we produce these facial expressions naturally and we don't have to learn them. It also suggests that we recognize them naturally too. So, groups of people all over the world, groups of people who have never come into contact with each other, are thought to express and understand emotions in the same way: by looking at facial cues, such as smiles and frowns. In the late 1960's researchers (e.g. Ekman and Friesen, 1967) argued that people used different facial expression as the main non-verbal way to communicate emotion.

The question for us is, can we communicate emotion through body movement as well as we can through the face? There are conflicting views about what emotions can be communicated and recognised from dance and body movements. Although Paul Ekman and Wallace Friesen suggested, in 1967, that facial cues are the main transmitters of emotion, more recent scientific dance-related research suggests that the moving body is just as powerful at signalling emotion as the face.

Matsumoto's 7 motives study

In 2004 a research team led by Mamiko Sakata, from Fukushima College in Japan, found that people could identify and recognise a wide range of emotions, such as happiness and loneliness, from a series of dance performances of Matsumoto's 7 Motives (Sakata, Shiba, Maiya & Tedenuma, 2004). Matsumoto's 7 Motives are representations of happy, sharp, dynamic, lonely, flowing, solemn and neutral feelings. The dances last between 4 and 34 seconds and are made up of high and low energy skips, turns, jumps and other dance movements. In Mamiko Sakata's study a single dancer, who maintained a neutral facial expression, performed the dances. People were then asked to watch the seven dances and to rate them according to their movement quality, their emotional content, and the extent to which they expressed six basic emotions (happiness, surprise, loneliness, fear, anger and disgust). The results of the study showed very clearly that people were able to recognise certain emotions and feelings through dance in Matsumoto's 7 Motives, without relying on facial expressions to provide additional information. For example, happiness was recognized in four of the seven dances, surprise and loneliness were recognized in two of the dances and anger was recognized in one of the dances. However, it was also found that people were not able to identify or recognise the emotions of fear or disgust in the dance performances. There may be two reasons for this. First, it might be the case that the dance performances did not communicate these emotions. Second, it might be the case that people are not able to recognise these emotions from body movement alone. Perhaps it's the case that people can recognise happiness, surprise, loneliness and anger from the moving body alone but to be able to recognize fear or disgust people need to see facial cues as well. Another study, this time using Hindu classical dance, helps us to clarify this.

Hindu classical dance & facial emotions study
A study by Paul Rozin, and his colleagues from the University of Pensylvania in the USA, looked at people's ability to recognize the portrayal of anger, disgust, fear and sadness in both Hindu classical dance, in which the face could not be seen, and in pictures of people making a facial expression of one of the emotions (Rozin, Taylor, Ross, Bennett & Hejmadi, 2005). Their results showed that people were able to recognise all of the emotions, including fear and disgust, through body movement alone. This suggests that in Sakata et al's study, where people were not able to recognize fear and disgust from Matsumoto's 7 Motives from body movement alone, it was not due to a limitation in the body's ability to communicate these emotions or of a limitation of the viewer to perceive these emotions from body movement cues. Nevertheless, Paul Rozin and his colleagues also found that fear and disgust were the hardest emotions to recognize generally.

Given that the New Zealand rugby team want to instil fear in their opponents by dancing the Haka before a game, and given how difficult it is for people to communicate and recognize fear from body movements alone, it's no surprise that the Haka uses several different cues in an attempt to amplify its fear-inducing signals. In addition to strong, stamping, tense body movements the Haka also includes eye-popping facial gestures and a loud, chanting chorus. These three cues, combined together, send the message loud and clear that their opponents have something to fear.

Physical cues in Recognising Emotion in Dance

There are, according to scientists, three types of physical cues that we use to recognise emotion in the moving body (Atkinson, Tunstall & Dittrich, 2007). These are structural or Form information, Kinematics and Dynamics.

Form
Form refers to the positions of the body or to the shape of the body movement, and how this changes over time.

Kinematics
Kinematics refers to velocity, acceleration and displacement

Dynamics
Dynamics refers to mass and the action of forces which produce changes in body movement.

The relative contribution of each of these cues to the recognition and interpretation of emotion has been examined in a series of studies, whereby form, kinematics and dynamics have been manipulated and controlled separately (e.g. Brownlow, Dixon,

Egbert & Radcliffe, 1997, Dittrich, Troscianko, Lea & Morgan, 1996, Walk & Homan, 1984).

The Point-Light Technique

Many of these studies have used a special lighting technique for examining people's ability to draw meaning from body movement and dance. Gunnar Johansson, a Swedish psychophysicist, developed this lighting technique (Johansson, 1973), which involved placing small lights, or reflective strips, onto people's joints (e.g. their wrists, elbows, knees, ankles etc.) before filming them as they performed basic actions, such as walking, cycling, gymnastics, climbing and dancing. Once these actions have been recorded onto video tape, all a person can see when they watch the video is a set of dots (points of lights) moving around the screen. This lighting technique has become known as the point-light technique. One of the really interesting things about the point-light technique is that when the dots are still, and not moving, they look like they make up a random pattern. When people look at them they have no idea that they are looking at points of light coming from a human body. However, as soon as the points of light start to move people know immediately, within a few hundred milliseconds, that the collection of lights represent the human form in movement, or biological motion as Johansson called it.

This is the case for both adults and children. In one study Pavlova, Krageloh-Mann, Sokolov & Birbaumer (2001) found that children from as young as 3 years were able to recognise human and non-human forms from moving point-light displays, but not from static point-light images. That is, people can recognize human movement from seeing just a dozen dots moving on a screen.

Johansson's point-light technique has been used in dance research to help us work out what it is about human movement that enables us to communicate emotion-based information through the body. Researchers, myself included, have filmed people dancing both in full-light and point-light conditions. The critical difference between filming dancers in these two different lighting conditions is that when dancers are filmed in full light the viewer is able to see the form of the dancer and they can see the quality of the movements, in terms of kinematic and dynamic information. However, when the dancers are filmed in point-light then the viewer is only able to see the kinematic and dynamic information and they are not able to see the form of the body. This technique allows us to ask whether the recognition of emotion from body movement comes from information about the form of the body or from kinematic and dynamic information. In one study carried out by Anthony Atkinson and his colleagues, (Atkinson, Dittrich, Gemmell & Young, 2004) participants were asked to identify the emotions of disgust, fear, anger, happiness and sadness from either full-light or point-light displays of moving and static bodies. They found that all the emotions were recognised at above-chance levels in both full-light and point-light conditions. This shows that even when you have minimal information about a moving body it is still possible to recognise emotions.

However, the researchers also found that not all emotions are recognised equally. Disgust, fear and anger were recognised significantly better in the full-light condition than in the point-light condition. This suggests that for disgust, fear and anger form makes a significant additional contribution to the communication of emotion over and above the contribution played by kinematic and dynamic information. For happiness and sadness there was no difference in levels of emotion recognition between full-light and point-light conditions, suggesting that for happiness and sadness form does not make an additional contribution to the recognition of emotion over and above that provided by kinematic and dynamic cues. The amount, and type, of information required to recognise emotions therefore varies depending on the emotion being expressed.

Particular areas of the brain are thought to be involved in the interpretation and recognition of emotion as it is expressed in the moving body. These areas are, in some cases, distinct from other areas of the brain, which are involved in the interpretation and recognition of emotion from facial cues.

Brain scan study

Peelen, Atkinson, Andersson and Vuilleumier (2007) carried out a brain scan study where they monitored activation in different parts of people's brain as they watched videos of emotional expression. They observed activity in two body-selective visual areas of the brain when people watch body movement representations of certain emotions. These areas are the extrastriate body area (EBA), which is in the lateral occipitotemporal cortex, and the fusiform body area (FBA), which is located in the fusiform gyrus, which is part of the temporal lobe and the occipital lobe. Peelen et al. scanned eighteen people, using a Magnetic Resonance Imaging (MRI) scanner. They showed participants full-light video clips of people expressing anger, disgust, fear, happiness, sadness and a neutral movement condition.

Peelen et al. found a relationship between the emotions that were being observed and the activation of different areas of the brain. They observed activation in both the extrastriate body area

(EBA) and the fusiform body area (FBA) when participants were watching videos of anger, disgust and happiness, and only the FBA became active when watching videos of fear. They also found that neither area showed increased activation when people were watching video representations of sadness.

Activation in the brain is hardly ever restricted to just one discrete region while we are processing information. This was certainly the case when Peelen et al. carried out their study as, in addition to seeing differential activation in the EBA and the FBA while people watched emotional displays, Peelen et al. also observed activity in an area of the brain called the amygdala. The amygdala is part of the limbic system and is located in the temporal lobes of the cerebral hemispheres. The activation in the amygdala corresponded with activity in the EBA and FBA. These findings suggest that certain clusters of neurones in different body-selective areas of the brain help us to identify and recognise emotion from the human body as it is moving. However, they do not offer the full story of how the brain processes different displays of emotion because, for one of the emotions presented (sadness) neither area of the brain under investigation became active, yet the emotion was still recognised. This leads to several conclusions and/or questions. First, and perhaps most obviously, is what is the neural basis of recognising sadness? What is the necessary pattern of neural activation that correlates with the recognition of sadness? It is clearly, according the findings of Peelen et al., not in the FBA, EBA or the amygdala. Secondly, perhaps the activation of the EBA, FBA and amygdala were secondary to the activation of other parts of the brain and, as such, were not necessary for the processing of the emotionally valenced visual stimuli despite their activation. A neuropsychological approach may help us to address the second question.

Cognitive neuropsychology

Cognitive neuropsychology is a branch of cognitive psychology that is based on the study of brain injured people. The basic logic of the approach is as follows. You start with a healthy brain which supports a whole range of normal cognitive functions (like language, memory, thinking and perception). Some damage happens to a certain part of the brain, for example, by either a blow to the head, surgery or a stroke, and this is related to a loss of a certain cognitive function, for example memory loss or an inability to use certain words. When this happens, and when scientists and medics can isolate the precise locus of the brain damage they are then able to draw a conclusion which suggests that the part of the brain that was damaged was responsible, in part, for processing the cognitive function that has been lost or attenuated.

The logic of cognitive neuropsychology can help us to understand whether particular areas of the brain (e.g. the amygdala) are responsible for processing the recognition of emotions from dynamic visual displays. In order to do this it would be necessary to test the emotion recognition of people who have localised damage to the amygdala with otherwise preserved neural functioning in other parts of the brain.

Although patients with this profile are relatively rare two such people (known in the medical literature as SM and AP) with localised damage to the amygdala were identified and they agreed to be tested. Atkinson, Heberlein and Adolphs (2007) tested SM and AP on their ability to recognise emotion from whole-body movement using both full-light and point-light videos of fear, anger, disgust, happiness, sadness and neutral emotional expression. Atkinson and his colleagues found that neither patient was impaired at emotion recognition (particularly the recognition of fear) compared to controls. This suggests that neuropsychological patients are able to recognise emotion from body movement without using the amygdala.

Specifically, Atkinson, Heberlein & Adolphs (2007) found that SM showed a deficit in identifying angry body movements, giving them a disgust label instead, perhaps suggesting that the amygdala plays a significant role in identifying anger. Furthermore, SM's responses to the point-light videos were less like the responses given by control participants (they were less accurate) than the responses she gave to the full-light videos. A tentative conclusion to this pattern of findings might be that the amygdala is not required to recognise emotion when form information is present in the stimuli (as is the case in full-light videos), but it plays a greater role in identifying emotion when only kinematic and dynamic information is present in the stimuli (as is the case in point-light videos). More research is clearly needed in this area to identify the neuropsychological processes involved in the recognition of emotion from body movement. However, it leaves us with the conclusion that despite the findings of Peelen et al., that activation in the amygdala (and the FBA and the EBA) might correlate with the recognition of certain emotional displays but they don't appear to be necessary. They might not be necessary for any number of reasons. Perhaps when they are damaged the functions of emotion recognition which they normally serve are accommodated in other brain regions, or perhaps their activation is an illusionary correlation. For example, when I get embarrassed I blush and my cheeks go red. However, if you removed my cheeks I would probably still get embarrassed, I just wouldn't be able to blush.

This research is informative and interesting, to a point. The major drawback of the work, from a dance psychology perspective, is that the portrayals of the different emotions were done using "acted" scenes instead of "dance". Basically, the researchers used actors to act out the movements for happy, sad, fear, anger and disgust. The movement patterns were short, iconic representations of the emotions. The best way to imagine what these movement patterns looked like is to imagine how a very bad actor, an actor who is over-acting in the local amateur dramatic society melodrama, might act out being angry or sad. Now, in your head, boil that image down to its essence and make the actor twice as bad as you originally thought. Once you've got that image in your head you are close to knowing what the viewers had to look at. Of course, it would have been very easy to recognize the emotional displays, as indeed the researchers found. Atkinson et al. (2004) found that the exaggeration of body movement by the actors enhanced recognition accuracy and also led to higher intensity ratings.

Communicating Emotion – an Artist's Approach.

I've always been entranced by watching dance when choreographers and dancers succeed in changing the way I feel. The first time I remember this happening was at a ballet performance of Romeo and Juliet (choreographed by Kenneth MacMillan). I was touched by the tenderness and love of Juliet and Romeo as they kissed on the bed. It was almost as if the couple weren't dancing, and I felt awkward, like a voyeur, spying on their intimacy. It was a beautiful portrayal and it touched me. It wasn't just that Romeo and Juliet is a passionate story. I had been completely unmoved, emotionally, by Frederick Ashton's ballet version of the same love story. I've no idea whether it was the choreography, the way the dance movements were expressed by particular dancers, or my own emotional state of mind (and heart) at the time that led me to be moved by MacMillan's version and not Ashton's. It was probably a combination of all of them.

Romeo and Juliet was also the inspiration for the musical West Side Story and the dancing in this film also made a big impact on me emotionally. The choreography, by Jerome Robbins, embodies and communicates perfectly the different emotional states of the Jets and Sharks. In one routine, danced to Stephen Sondheim's "Cool", we can feel the pent-up frustration in the tight, sharp, staccato movements and the central character's anger is portrayed by the explosive physical release of muscles and limbs, which is echoed by the movements of the supporting dancers. All this is performed while the dancers are trying to maintain a mask of cool relaxation. This is a great example of how the purely scientific research approach falls short of capturing the essence of real emotional expression and communication. In the experiments described above the researchers investigated how single, iconic emotions (e.g. happiness, sadness, anger and so on) were portrayed in movement and recognized. In real choreographed representations of emotional states, it is extremely rare to find the representation of a single emotion. Usually, emotional representations are of complexes of emotional states. For example, when people are feeling acutely happy, they may also experience simultaneous feelings of guilt, or when people are feeling proud they may also have a sense of embarrassment. This is not to say that these pairs of emotions always go together, just that it is more likely that people will feel "sets" of emotions at the same time rather than single isolated emotions. It also suggests that people might also be able to recognise emotion-complexes in other people too.

"23 Feelings in Dance" project

I wanted to know how dance makers, choreographers, would go about portraying sets of emotions in dance and I wanted to know if audiences watching dance could recognize potentially complex sets of emotions. In the spring of 2010 I put out a call to choreographers to create a three-minute piece of dance based on a set of emotions drawn from a list of 190 feelings-based words. Some of the words in the list were obvious, feelings-based words like love, afraid and proud, that are easy to recognize in people's movements, but the majority of the words in the list were subtler, like benevolent, estranged and mixed-up, that might be harder to portray physically and recognize in the way people move. Choreographers responded brilliantly to the brief and I was sent hundreds of dance films. I received entries in jazz dance, tap, ballet, and lots of contemporary dance pieces that portrayed complex sets of emotions. The choreographers often sent me a written explanation of their work but I always watched the pieces first before I read any of these written explanations.

I was preparing a show for the 2010 Edinburgh Festival Fringe called "Dance Doctor, Dance! The Psychology of Dance Show" where I would discuss the science of dance, so I decided to use the 23 individual performances of my show to see whether audiences could recognize the emotions and feelings expressed in the pieces created by the choreographers. From the hundreds of entries I received, I selected 23 pieces based on the complexity of the emotions expressed. Therefore, I didn't select any pieces where the choreographer was trying to express just a single, iconic emotion, like happy or sad and I avoided pieces where the form of dance expression used a vocabulary of known movement gestures to communicate an idea (like in some forms of classical dance). What I was left with was a collection of works that were mainly derived from contemporary dance practice. I then invited each of the choreographers to come to Edinburgh and perform their three-minute piece on one of the nights of my 23-night run. I called this collection of work 23 Feelings in Dance.

My aim in 23 Feelings in Dance was to leave the constraints of a scientific laboratory behind and see if audiences could recognize a choreographer's intention to portray emotion and feelings through dance. Each night I spoke about the science of emotion recognition and then I introduced one of the 23 Feelings in Dance pieces. I didn't tell the audience what to expect. Once the piece was over I asked the audience directly what emotion or feelings they thought were being represented in the piece. The audience would shout things out and we would discuss what they saw and then I would bring the choreographer onto the stage to discuss the way in which they started to work with dancers to represent certain emotions or feelings. What struck me, night after night, was that the audiences were spot on in their understanding of the pieces even when abstract forms of dance were used and when very complex patterns of emotion were represented. Sometimes the audience found it difficult to articulate what they'd seen, but they still accurately described the essence of the choreographer's intention. Let me give you an example.

One of the pieces was a solo, choreographed by a woman and danced by a man. This is relevant. The piece was about what it felt like, for a woman, to lose an unborn child during pregnancy. When, after the piece, I asked the audience to tell me what they thought the piece was about there was silence. I looked around for a minute until someone said that they didn't know, but that they thought it was something to do with loss. As they said this they made a downward pushing motion with their hands, moving their hands from their chest down to their lap. This movement had not been used in the dance piece. They made this downward pushing motion a couple of times as they said, "something to do with loss". Another member of the audience said they thought it was based on "decay, or dying inside", but that they couldn't place this within a meaningful context. This was interesting to me because it showed how it is possible to communicate feelings and ideas through movement that people understand implicitly, without necessarily understanding the context of the piece explicitly. In this piece, the male dancer was expressing what it felt like for a woman to lose a pregnancy and yet despite the logical inconsistency of this piece being danced by a man it was still possible for the audience to understand its emotional essence.

Summary

Combining science and dance as art gives us a rich picture of how we communicate emotions through dance and body movement. It is clear from the scientific research that single iconic emotions can be communicated effectively even when the visual stimuli do not show the form or body of the dancer, whilst a dance-as-art interpretation shows that complex patterns of emotions can be expressed and understood through the medium of dance.

The communication of emotion is a fundamental part of human communication. Emotion recognition is, or perhaps was, vital for human survival, social bonding and the forming of intimate interpersonal relations. Perhaps our ability to communicate and understand human emotions through body movement should be considered alongside the other aspects of innateness discussed in Chapter 1, and, as such, might provide further support for the idea that we are born to dance.

EPILOGUE

From Dancer to Dance Psychologist

I was asked by Rhod Gilbert, the comedian, to provide an answer to the question "What's the point of dancing?" It was for Rhod's TV show. While I was giving my answer Rhod asked me whether it would be better to have the legs of John Travolta (the disco-dancing legend) or his brain, in order to be a great dancer. I said the answer was easy. It was the brain. If you had Travolta's brain, you too could dance like Travolta. If you only had his legs, I thought, you'd dance like a giraffe.

Thinking about whether it is the body or the brain (or the mind) that is most important for dance is an interesting distinction. I was a dancer before I was a psychologist and, as a dancer, I would have certainly defended the primacy of the body over the mind when it came to dancing. As a young dancer I "felt" what it was like to dance in my "body" and it was my "body" that I thought I was training when I danced. It was my "body" that connected with the rhythm of the music when I danced in my bedroom and in nightclubs and it was my "body" that I looked at when I danced in front of a mirror. I didn't need to think when I was dancing, my body seemed to move naturally. Surely, I would have thought, we dance with our body. Don't we switch off our mind when we dance? Dancing helps us to lose our inhibitions, it helps us to forget our daily worries, and dancing releases chemicals in the brain that help us to relax, to inhibit activity in the brain – in one sense, to switch itself off. However, the more I've learnt about the science of dance the more I am convinced that the mind is at least as important for great dancing as the body. Not only does the mind help us to be better dancers it can also make dancing incredibly difficult for us. When people come to me and ask me to help them become "better", more natural, dancers, I always start with their state of mind and then I work on breaking down the mind-blocks that are preventing them from dancing effortlessly, easily and naturally. In this book, I tell you about dancing bodies and dancing brains. To put it all into context I'm going to tell you about my own dancing body and how I eventually engaged my dancing mind.

I have danced all my life. I don't remember the precise moment I started to dance, but I do remember that dancing was something that just happened. I felt compelled to move. Compelled to dance.

Early Childhood Dance Memories

As a child, I loved to dance at home, at parties and at social gatherings. My parents both worked at a hospital where there was a thriving staff social club. This was a great place to dance. Bands played frequently and there were disco nights (it was, after all, the 1970's). It seemed to me that by late in the evening everyone was dancing - men, women, old and young. During the same period we used to go on family holidays to holiday camps where, again, dancing was part of the fabric of the holiday. The social dancing at holiday camps was also inclusive and varied. There'd be freestyle disco dancing, line dancing, country dancing, ballroom dancing and Ceilidh dancing. Everyone danced. Dancing with my mum, dad and sister, I learnt the St Bernard's Waltz, the Gay Gordons, the Slosh and a little dance routine made famous by the pop group Brotherhood of Man, to a song called "Save Your Kisses For Me".

As a young teenager, I used to dance myself into a euphoric state. I would play records in my bedroom and dance and dance and dance until my head buzzed. Nothing else gave me the rush I got from dance. My bedroom window overlooked a busy road. At night, I would open the curtains, turn on all the lights and dance in front of the window-turned-mirror for hours at a time. I was completely unaware that everyone outside could see me. From outside people would have seen a sweaty teenage boy bouncing frantically around the room. I must have looked ridiculous, but I didn't care, because I felt fabulous.

By the time I was 14 I'd started to have ballet lessons with a local teacher who had a dance studio in the basement of her house. My teacher was a grey-haired woman with grown-up children. Helen had been a student at the Royal Ballet School and she told me that she'd had to give up the dream of being a professional ballet dancer because she'd become pregnant at 18. She told me that going to see ballet was very difficult for her, because she longed to be on the stage, and seeing others live her dream upset her too much. She would tell me stories like this, about her life as a trainee dancer and of her love for dance and of her regrets, after our class, as we sat in her sitting room listening to recordings of ballet music. This was the first time I'd heard someone talk about their passion for dance, where that passion had left a permanent scar. I'd experienced the joy that dancing can bring, but Helen's life seemed to be defined by both the ecstasy and the agony of dance; the love of dance which had been fine-tuned through thousands of hours of dedicated practice and then the lost love followed by regret and mourning. Perhaps it was at that moment that the seeds of Dance Psychology were sown in my soul.

Secondary School

School and dancing didn't mix well in the 1970's. One day, I knew, I wanted to earn my living as either a dancer or a choreographer in theatres and when I was 15 I had to see my school careers advisor to arrange a week of work experience. I told her I wanted to work in a local theatre. I'd be happy to do anything, working in the box office, front of house, backstage. She thought my request was ludicrous and told me that such a job would be a waste of time. "I've got a position for you," she said, "as a sheet metal panel beater at a local engineering firm" (this was the standard destination at the time for boys, like me, who were unlikely to achieve many formal school qualifications). I swore. My father was called and we all ended up in the Headmaster's office. The Headmaster raised his voice, we listened. My dad was a sensible man. He pointed out that it did seem rather silly to send a boy with a passion for dance and theatre and no interest in engineering to work as a panel beater. The Headmaster suggested I audition for a place on a Theatre and Creative Arts course at a college on the other side of the county. We had no idea such courses existed. The careers advisor sat silently, nose out of joint.

Remedial English

I spent two years studying Theatre and Creative Arts at East Herts College. It was a wonderful time. I danced, sang, improvised, acted, choreographed, laughed, made papier mâché masks and learnt how to apply stage make-up. There was an expectation that we should also complete some traditional academic work. We had lessons in theatre history and, because I had left school without an English O Level, I was expected to do it at the college. I tried and failed it two or three times. This wasn't a surprise. At secondary school I had been placed in a remedial English class. On one side of me there had been a boy who couldn't speak English - they were teaching him basic vocabulary - and on the other side of me there was a boy who couldn't write the letters of the alphabet - they were teaching him to type on an old-fashioned clunky typewriter. I could speak English and I could write the letters of the alphabet, I just wasn't cut out to work with the written word. Words didn't always make sense to me. I could read simple words and strings of unambiguous words but I had three reading difficulties that made advanced reading and comprehension difficult.

Reading Difficulties

My first difficulty was with reading, or "sounding out", words that have irregular spelling. For example, in the sentence, "I didn't choose to lose my shoes" I'd spend a long time looking at the words and trying to sound them out as if choose, lose and shoes had distinct sounds, as well as distinct letter groups, or I'd write the words down as if they all had the same letters, such as "I didn't chose to lose my shose" or "I didn't choose to loose my shoose".

Reading and writing words was very messy. Even now I feel a slight terror as I look at these words and wonder how anyone ever learns how to read, and I still have largely illegible handwriting, which masks my inability to spell correctly. Had I been at school today I may have been diagnosed with dyslexia.

My second difficulty was with reading sentences that had long-distant dependencies (yes, I've learnt the lingo now). An example of a sentence with a long-distance dependency is "The car, parked next to the bank, which was robbed yesterday by a man wearing a balaclava, was blue". In this sentence, it is the car that is blue. But I had a very poor short-term memory for words and when I read ("read" -- another ambiguous word!) sentences like this I wouldn't know whether it was the car, the bank or the balaclava that was blue. I had extreme difficulty in analysing (or parsing) the structure of sentences and that made reading complex text for meaning almost impossible.

My third difficulty was a visual one. When I looked at blocks of printed text on the page my vision seemed to go out of focus and all I saw in each paragraph was a single black block of ink. I could break through it and see the individual words but it was a huge effort and very tiring. Handwriting was easier to read, as was poetry. I found poetry, play texts and scripts easier to read than text books because they typically had more white space, which made it easier for me to focus on individual words. I could read some poetry, even some Shakespeare, because the rhythm of the words gives a clue as to where to look for meaning and a rhyming structure tells you explicitly how to sound out some words. Nevertheless, I survived without an O level in English and I moved on from East Herts College to study dance and musical theatre at the Guildford School of Acting. I eventually overcame these reading difficulties and I believe that dancing and finding a rhythm in words helped. I describe how this happened later.

From College to Drama School

The aim of most of the people studying Theatre and Creative Arts at East Herts College was to get a place at one of the major dance or drama academies. We spent weeks preparing multiple audition pieces (songs and monologues) and getting our technical dance skills in jazz, tap and ballet up to standard. We all worked hard and believed we had what it took to win our place. The audition process was strict and stressful. Because there was a lot of competition it was certainly the case that not everyone who was good enough actually would win a place. Only about a third of us got a place. I went to the Guildford School of Acting, and friends went to RADA, Central School of Speech & Drama, Webber Douglas, London Studio Centre and Rose Bruford College. But if we thought that auditioning to get in to drama school was tough then we knew nothing of how stressful it would be when we had to audition to get out of drama school.

The Audition Process

After three years of being trained, shaped, pumped up and polished at drama school, which I loved and loathed in equal measure, we were released through the glittering shimmer-curtain out into showbiz land. The only thing that stood between us and a job was the most brutal process of all, the audition. An audition is where you have to showcase your talents by singing, dancing and acting in front of a panel of choreographers, directors, producers and musical directors.

It's exactly the kind of audition process that makes The X Factor compulsive viewing. At an open call, there could be hundreds of people dancing and singing to get noticed. Lots of the women would look the same as each other (no surprise really, considering they were all trained, shaped, pumped up and polished at similar drama schools) and I would sometimes feel like I was standing in a room full of mannequin dolls. The audition process was vile - until you got the job. Then it was suddenly a fantastically fair process designed to allow talent to get noticed. When you get a job it seems that natural justice has prevailed. When you don't get a job, the rejection can chip away at your self-confidence, and repeated rejection can chip, chip away until your spirit is crushed. Some researchers have found that only about 3 people in every 100 that graduate from a vocational dance school - dance schools where people had to audition to get in - ever earn their living as a dancer.

Professional Dancer

I was very lucky to work as a professional dancer for several years. My first job out of GSA was with George Mitchell's Minstrel Show. This was a variety show that toured throughout England and Scotland and played at some of the biggest regional theatres, including the Nottingham Theatre Royal, the Birmingham Hippodrome and the Sunderland Empire. We even played at the end of a couple of piers. I continued to work as a professional dancer, and at one point I worked for a UK-based choreographer and went off on a tour of the Caribbean on a cruise liner. This was an experience, a character-building one. I was underpaid for several months, then I was put off the ship in the Dominican Republic without a ticket home or any money, and then the ship sank. The first two situations were connected to my behaviour, but I had nothing to do with the sinking of the ship.

Learning to Read

The beginning of the end of my initial professional performing career started while I was performing in Aladdin, the pantomime, at the Richmond Theatre in London. In the summer before the pantomime season started I met a group of very bookish people. They were all high flyers at prestigious universities. I think they were the first real intellectuals I'd met, or at least they were the first people of my own age I recognized as such. The group was made up of girls and boys who were confident and articulate. They all came across as well read, knowledgeable, worldly-wise and, most of all, supremely confident. They gave the impression that they could do anything they set their minds to. Although I only spent a small amount of time with the members of this group they had a profound effect on my life.

Frank was the alpha-male of the group. He'd been head boy at a famous public school and was studying literature at Oxford University when I met him. His parents were wealthy in a way I couldn't comprehend and his life was the polar opposite of mine, in every way. I thoroughly enjoyed Frank's company. He taught me how to play bridge and he was quick-witted. Even though Frank knew about world affairs, politics and fine art, and he had a school satchel full of qualifications, I felt comfortable in his company. I had my successes in dance and performing arts and he had his as a by-product of a traditional education. Everything was fine until Frank crossed over into my world. Back at Oxford, he put on a production of Mike Leigh's Abigail's Party and invited me to the performance in his college. I went along expecting it to be rubbish. But of course, it wasn't. He'd done a really good job. I left Oxford the next day feeling cross, cheated and humiliated. I kept thinking how unfair it was that despite his ability to achieve in one area he was also able to achieve in another separate area, my area. I'd felt that I was Frank's opposite but equal. Yet his directorial achievement put everything out of balance. Basically put, I thought he could do what I could do but I couldn't do what he did.

This thought stayed with me for weeks, going around my head. It niggled and bothered me. I started to think how unfair it was that some people could, seemingly, do everything, and others of us, me, couldn't. It bothered me that I couldn't read properly. It seemed to me that the only real difference between Frank's group of bookish friends and me was that they spent their lives in books and written information and I didn't, couldn't. It bothered me that I didn't know stuff because I couldn't get anything other than the most basic information from the written word, that I'd never read a "classic" book. In fact I'd never read a book from cover to cover. I felt I was missing out on great conversations about things I knew nothing about because I wasn't reading.

Most of all it bothered me that Frank could do what I could do but I couldn't do what he did. So I changed the situation. I decided to do what Frank did. I knew I could dance. Other people knew I could dance. I wanted to read Russian literature in translation. So that's what I did. And it changed everything.

Of course, I had my three reading difficulties but I was so "bothered" that I wasn't going to let these get in my way. I warmed myself up for my reading marathon by reading The Cross and the Switchblade by David Wilkerson. I chose that because one of the plain yet pretty girls in the group had just read it in a couple of days. I decided to approach reading as though I was learning a new dance. I approached it as if it were a problem-solving exercise. I knew that I wouldn't understand every sentence, or even every word, but I thought if I could find a rhythm, or sets of different rhythms, within the story then I could break it down and learn it "one eight at a time". The other things I knew from being a dancer was that perfection comes from relentless hours of practice and success comes in incremental steps.

So, I started to read. Whenever there was a word that I didn't know how to sound out I simply made up a sound for that word. There was no one to tell me off if I sounded out "yacht" as anything other than a word that rhymed with "got". The closest I came to sounding out "yacht" was to put a "y" sound in front of "act" as 'y-act". I couldn't make sense of the "h". I still can't see the logic of how to sound out "-acht". I felt that if the arrangements of letters in words could be arbitrary then the sound in my head could be arbitrary too. Whenever there was a long sentence with long-distance dependencies ("the car parked by the bank…was blue"). I simply broke it down and learnt it 'one eight at a time', and whenever the page went black I used to close my eyes, relax, and open them again and try to focus on the shape of one of the words. All this took a very long time. I read The Cross and the Switchblade for about 12 hours a day for two weeks. When I'd finished it I had no more than a basic understanding of the story, but I knew then that I could read a book and I was one step closer to doing what Frank did. I guess I'd also learnt something from the essence of The Cross and the Switchblade, that sometimes you have to take on challenges that at first seem impossible. My next book was Tolstoy's Anna Karenina and then Resurrection and then I moved on to Dostoevsky and Turgenev. For light relief I read Jeffrey Archer. I was like an ocean-going liner; once I'd started I just kept going, moving slowly from one book to another (and, unlike the MV Oceanos, my Caribbean cruise liner, I tried not to sink).

By the time I arrived at the rehearsal room for the pantomime season of Aladdin in late autumn I knew things were changing. Knowing that I could read a book meant that I could read any book. Knowing that books were gateways into whole new worlds meant that I could go into any new world that I wanted. Knowing also that Russian literature was an academic subject, studied by clever people at Oxford University, and knowing that I could read Russian literature meant, in my naive head, that I too could study something, anything, at a university. This possibility was overwhelming. When the panto season ended at the end of January I set out on a new path. I left my life in London having decided to start an academic journey. Despite the fact that I had no qualifications, except for a CSE grade 1 in drama and a diploma from the Guildford School of Acting, I had read Wilkerson, Tolstoy and Jeffrey Archer and in the worlds of their books anything is possible.

Entering Academia

In the real, non-fiction, world getting into university was harder than I thought it would be. I decided to study psychology. I thought that I could combine psychology with drama and dance and perhaps train as a drama or dance therapist and use the creative arts to help people. I bought and read books by Freud and Jung and I read case studies, such as Dibs - In Search of Self, Cry Hard and Swim and The Man who Mistook his Wife for a Hat. Although all of these books were found on the psychology shelf of my local bookshop it wasn't clear to me, once I'd read them, how they related to each other. The Man who Mistook his Wife for a Hat was about neurological case studies, Dibs - In

Search of Self was about a series of play therapy sessions with a young boy with a high IQ, who had been misdiagnosed with autism, and Cry Hard and Swim was the story of an incest survivor. As I read Freud and Jung I was expecting an explanation of these different aspects of psychology or some answers to the questions raised in the other books. I found neither explanations nor answers in the work of Freud or Jung. At the same time I started to contact psychology departments at a couple of universities to ask if they'd let me onto their course. I had no idea how to go about applying to university but I was told by the psychology departments that I would need at least one A level before I could apply.

The Dance Break

I enrolled in an evening class to study for an A level in Psychology. To do this meant that I had to be stable. I had to be based in one place for a year and I wouldn't be able to work in the evenings. This set of circumstances meant that I couldn't work as a professional dancer while I was studying. This was hard. I'd made a decision to give up the only thing I was good at to do something that I wasn't very good at. I took odd jobs as a delivery driver while I started to study psychology. I found the subject matter interesting and I thoroughly enjoyed reading about the different areas of psychology and the methods used to collect and analyse data. I attended classes twice a week while I made formal applications for a place at University. It was great. I'd have just one year of doing odd jobs before I'd be a full-time university undergraduate. Everything was fine, until I was rejected by all five of the universities to which I had applied. Manchester, Sheffield, UCL, Bristol & Durham all said no. This came as a blow. A hard one. It was winter time when the rejections arrived, over a period of several weeks. By the time the fifth rejection arrived I was ready to run. I felt extremely despondent and so I stopped going to the evening-class course and wondered what on earth I'd done. I'd given up a career as a dancer, the only thing I was good at, the only thing that felt entirely natural to me, to go to University and I hadn't got a place. What an idiot.

I didn't read any psychology text-books for three or four months. I couldn't face it. At the time, I was dating a girl (Lindsey) who I'd met at the A level evening class. She too was going through a career change and needed her A level in Psychology to go to university. She'd been accepted by the university of her choice and was sailing through the course. As she approached her end-of-year exams she started to nag me. I'd missed more than half of the course but she nagged and nagged and nagged until I agreed to return for the last few classes of the year and sit the A level exam. Lindsey and I worked together every night for a month on the course content. It was the best thing I'd ever done for two reasons. First, I sat the exam, and passed (just, but it was enough to keep me in the academic game) and second, I married Lindsey a few months later and she's been my life partner ever since.

After scraping a pass in A level psychology another year of uncertainty followed. As Lindsey was settling into her university course I took a job as a trainee bus driver in London. Driving a bus out of Victoria Bus Station at 5am on a cold, wet morning is a long way from dancing in a costume of pink crushed silk on a cruise ship in the Caribbean, in so many ways. I was missing the dance studio and was in a fairly dark place. I decided to give university one more attempt and if that didn't work then I would go back to theatre. This time I thought I should brush up on my academic English skills and as I'd learnt to enjoy reading I thought

I'd try to study for an A level in English. I enrolled on a correspondence course. I was doing shift work at the bus depot and studying for the rest of the time. I also took a more strategic approach to getting a university place. I went to visit several university psychology departments and met with various admission tutors. This time one of the colleges (Roehampton Institute – now Roehampton University) made me an unconditional offer to study for a BSc in Psychology and English. I accepted the offer, turned a pirouette, and returned my bus driver's uniform.

I formally started my university journey at Froebel College, Roehampton, in September 1990. The grounds of Froebel College were beautiful and it felt as though I had been picked up and dropped back in time. During my three years there I became fascinated by neurobiology and neuropsychology, which concern the study of the biological make-up of the brain and what happens to people when the brain is damaged. I also danced a lot and became the president of the Trapdoor Theatre Company, the college drama society. I produced and directed a production of Lysistrata, which we took to the Edinburgh Festival Fringe, I danced in countless pieces for the undergraduate dance students and, for a few months, I trained and danced with the Chelsea Ballet Company. At the time psychology and dance were very separate parts of my life. I learnt about psychology in the lab and I danced in the studio. I'd also decided that I didn't want to train as a dance therapist (I was far too excited by the study of the brain and its disorders) and so, I thought, I would have to keep dance and psychology separate.

I graduated from Roehampton, with a University of Surrey degree, in 1993 and took up a scholarship to study for an MSc in Neural Computation at the Centre for Cognitive and Computational Neurosciences at the University of Stirling. Neural

Computation is all about how we build models of the working brain using mathematics and artificial networks. We were a small, odd, cohort of students made up of computer scientists, physicists, mathematicians and psychologists and our aim was to understand how we could build plausible models of the brain and then inflict "brain damage" on them so that we could see how they recovered. It was an ambitious target. I found the mathematics baffling at first. Lectures would consist of slide after slide of mathematical formulas, made up of what seemed like hundreds of Greek symbols. Needless to say, my evenings were spent learning to distinguish and name the symbols for theta, delta, lambda (which I thought was a drama school).

I graduated from the University of Stirling and moved on to the University of Essex to take up a scholarship to study for a doctorate in experimental cognitive psychology. Cognitive psychology is about how humans think, learn, solve problems, use language, perceive the world and remember. Experimental cognitive psychology involves a great deal of laboratory work, testing, for example, the limits of memory and language under carefully controlled conditions. I spent three years in a very small lab measuring how much time it took people to read lists of words and afterwards to see which words were the easiest to remember. I was trying to understand how humans learn and remember. My original aim in doing this work was to have an understanding of the mental structure of memory and language processing, so that when we met people with brain damage, who had damage to their memory and language systems, we could help them, and clinicians, to develop appropriate programmes of rehabilitation. I was fascinated by learning and memory, particularly after my own experience of coming late to learning. I completed my doctorate and took up a post-doctoral position in the Research Centre for

English and Applied Linguistics, in the Faculty of English at Cambridge University.

At Cambridge University, I worked as a psychologist on a project to examine how people learn more than one language. I was interested in how people "think" in different languages and about how they store and remember lots of words that might have the same or different meanings across different languages. I was interested in how people read new (foreign) words and how they make sense of them, and then learn to understand complex linguistic patterns. I also became interested in the relationship between dyslexia and memory. For the first time, I started to read about some of the problems that people with dyslexia have seeing, coding and remembering words. The descriptions I read in the literature could have been written about me and the problems I had with reading. The difficulty of reading "exception" words, that is words whose sound doesn't match their letters, the difficulty of remembering which words in a very long sentence go together and the difficulty of focusing on individual words in large blocks of text. Learning about the cognitive models of these reading difficulties helped me to understand why I had found reading for meaning so difficult and how I had been able to overcome these difficulties.

From Remedial English to Cambridge University

My time at Cambridge University, as an academic psychologist, marked the end of a long journey. It had taken me ten years to go from being a dancer who'd never read a book to being a Cambridge academic, and now I could say with some confidence that I could do what Frank did. I had a BSc, MSc, and PhD in psychology, I'd been awarded two major scholarships and I had an academic post at one of the best, most competitive, universities in the world. I felt as though I'd reached the summit of Mount Everest. So, I did exactly what everyone does when they reach the summit of a mountain; I turned around and walked back down. I left Cambridge University and started to plan how I could combine my expertise in psychology with the subject I loved most in the world, dance.

I set up the Dance Psychology Lab.

REFERENCES

Adiputra N., Alex P., Sutjana D.P., Tirtayasa K., Manuaba A. (1996). Balinese dance exercises improve the maximum aerobic capacity. Journal of Human Ergology (Tokyo) 25(1), 25-29.

Ainsworth, B. E., Haskell, W. L., Whitt, M. C., Irwin, M. L., Swartz, A. M., Strath, S. J., O'Brien, W. L., Bassett, D. R. Jr., Schmitz, K. H., Emplaincourt, P. O., Jacobs, D. R. Jr., & Leon, A. S. (2000). Compendium of physical activities: an update of activity codes and MET intensities. Medicine and Science in Sports and Exercise, 32(9), S498-504.

Atkinson, A. P., Dittrich, W. H., Gemmell, A. J. & Young, A. W. (2004). Emotion perception from dynamic and static body expressions in point-light and full-light displays. Perception, 33, 717-746.

Atkinson, A. P., Heberlein, A. S. & Adolphs, R. (2007). Spared ability to recognise fear from static and moving whole-body cues following bilateral amygdala damage. Neuropsychologia, 45, 2772-2782.

Atkinson, A. P., Tunstall, M. L. & Dittrich, W. H. (2007). Evidence for distinct contributions of form and motion information to the recognition of emotions from body gestures. Cognition, 104, 59-72.

Atkinson, R. C. & Shiffrin, R. M. (1968). Human memory: A proposed system and its control processes. In K. W. Spence &

J. T. Spense (Eds.), The Psychology of Learning and Motivation, Vol. 2. London Academic Press.

Baddeley, A. D. (1986). Working Memory. Oxford: Clarendon Press.

Bakker, F. C. (1988). Personality differences between young dancers and non-dancers. Personality and Individual Differences, 9(1), 121-131.

Bartlett, F.C. (1932). Remembering: A Study in Experimental and Social Psychology. Cambridge University Press.

Baumeister, R. F., & Leary, M. R. (1995). The need to belong: Desire for interpersonal attachment as a fundamental human motivation. Psychological Bulletin, 117(3), 497–529.

Beaulac, J., Olavarria, M., & Kristjansson, E. (2010). A Community-Based Hip-Hop Dance Program for Youth in a Disadvantaged Community in Ottawa: Implementation Findings. Health Promotion Practice, 11(1), 61S-69S.

Belardinelli, R., Lacalaprice, F., Ventrella, C., Volpe, L. & Faccenda, E. (2008). Waltz Dancing in Patients with Chronic Heart Failure: New Form of Exercise Training, Circulation Heart Failure, 1, 107-114.

Bellieni, C. V., Cordelli, D. M., Bagnoli, F. & Buonocore, G. (2004). 11- to 15-year-old children of women who danced during their pregnancy. Biology of the Neonate, 86(1), 63-65.

Ben-Tovim, D. I., & Walker, M. K. (1991). The development of the Ben-Tovim and Walker Body Attitudes Questionnaire (BAQ), a new measure of women's attitudes towards their own bodies. Psychological Medicine, 21, 775–784.

Blackman, L., Hunter, G., Hilyer, J. & Harrison, P. (1988). The effects of dance team participation on female adolescent physical fitness and self-concept. Adolescence, 23 (90), 437-448.

Blanchette, D. M., Ramocki, S. P., O'del, J. N., & Casey, M. S. (2005). Aerobic exercise and cognitive creativity: Immediate and residual effects. Creativity Research Journal, 17(2&3), 257-264.

Brown, S., Martinez, M. J. & Parsons, L. M. (2006). The Neural Basis of Human Dance. Cerebral Cortex, 16, 1157-1167.

Brown, W. M., Cronk, L., Grochow, K., Jacobson, A., Liu, C. K., Popovic, Z. & Trivers, R. (2005). Dance reveals symmetry especially in young men. Nature, 438, 1148-1150.

Brownlow, S., Dixon, A. R., Egbert, C. A., & Radcliffe, R. D. (1997). Perception of movement and dancer characteristics from point-light displays of dance. Psychological Record, 47, 411–421.

Burgess, G., Grogan, S. & Burwitz, L. (2006). Effects of a 6-week aerobic dance intervention on body image and physical self-perceptions in adolescent girls. Body Image, 3 (1), 57-66.

Burkhardt, J. & Brennan, C. (2012). The effects of recreational dance interventions on the health and well-being of children and young people: A systematic review. 4(2), 148-161.

Burt, R. (1995). The Male Dancer: Bodies, spectacles, sexualities. Routledge: London & New York.

Campion, M. & Levita, L. (2014). Enhancing positive affect and divergent thinking abilities: Play some music and dance. The Journal of Positive Psychology, 9(2), 137-145.

Carney, D. R., Cuddy, A. J. C. & Yap, A. J. (2010). Power posing: Brief nonverbal displays affect neuroendocrine levels and risk tolerance. Psychological Science, 21(10), 1363-1368.

Carvalheiro, S. & Rodrigues, L. X. (2009). Memory span in dance: Influence of age and experience. International Symposium on Performing Science, 179-184.

Causley, M. (1980). Introduction to Benesh Movement Notation (Dance). Ayer Co. Publishing.

Chang, Y. K., Labban, J. D., Gapin, J. I. & Etnier, J. L. (2012). The effects of acute exercise on cognitive performance: A meta-analysis. Brain Research, 1453, 87-101.

Cirelli, L. K., Einarson, K. M. & Trainor, L. J. (2014). Interpersonal synchrony increases prosocial behaviour in infants. Developmental Science, 17 (6), 1003-1011.

Coopersmith, S. (1968). The antecedents of self-esteem. San Francisco: Freeman.

Coupland, J. (2013). Dance, ageing and the mirror: Negotiating watchability. Discourse and Communication, 7(1), 3-24.

Crotts, D., Thompson, B., Nahom, M., Ryan, S. & Newton, R. A. (1996). Balance Abilities of professional dancers on select balance tests. Journal of Orthopaedic & Sports Physical Therapy. 23 (1), 12-17.

Darwin, C. (1872). The expression of emotion in man and animals. London: John Murray.

Di Blasio, A., De Sanctis, M., Gallina, S., & Ripari, P. (2009). Are physiological characteristics of Caribbean dance useful for health? The Journal of sports medicine and physical fitness, 49(1), 30-34.

Dittrich, W. H., Troscianko, T., Lea, S. E. G., & Morgan, D. (1996). Perception of emotion from dynamic pointlight displays represented in dance. Perception, 25, 727–738.

Dooling, D. J. & Lachman, R. (1971). Effects of comprehension on retention of prose. Journal of Experimental Psychology, 88, 216-222.

Eerola, T., Luck, G., & Toiviainen, P. (2006). An investigation of preschoolers' corporeal synchronization with music. In M. Baroni, A. R. Addessi, R. Caterina, & M. Costa (Eds.), Proceedings of the 9th International Conference on Music Perception and Cognition (CD-ROM (pp. 472–476). Bologna, Italy: Università di Bologna.

Eisenmann, J. C., Katzmarzyk, P. T., Perusse, L., Tremblay, A., Despres, J. P. & Bouchard, C. (2005). Aerobic fitness, body mass index, and CVD risk factors among adolescents: the Quebec family study. International Journal of Obesity, 29(9), 1077-1083.

Ekman, P. & Friesen, W. (1967). Head and body cues in the judgment of emotion: A reformulation. Perceptual and Motor Skills, 24, 711-724.

Eyigor, S., Karapolat, H., Durmaz, B., Ibisoglu, U., & Cakir, S. (2009). A randomized controlled trial of Turkish folklore dance on the physical performance, balance, depression and quality of life in older women. Archives of Gerontology and Geriatrics, 48, 84-88.

Farnell, B. (1999). Moving bodies, acting selves. Annual Review of Anthropology, 28, 341-373.

Fink, B., Seydel, H., Manning, J, T. & P. M. Kappeler (2007). A preliminary investigation of the associations between digit ratio and women's perception of men's dance. Personality and Individual Differences, 42, 381-390.

Fischer, S., Hallschmid, M., Elsner, A. L. & Born, J. (2002). Sleep forms memory for finger skills. Proceedings of the National Academy of Sciences, 99(18), 11987-11991.

Fisher, J. & Shay, A. (2009). When Men Dance: Choreographing masculinities across borders. Oxford University Press.

Flores, R., (1995). Dance for health: improving fitness in African American and Hispanic adolescents. Public Health Reports, 110(2), 189-193.

Fredrickson, B. L. (2004). The broaden-and-build theory of positive emotions. Philosophical Transactions of the Royal Society of London, Series B, 359, 1367-1377.

Gardner, S. M., Komesaroff, P. & Fensham, R. (2008). Dancing beyond exercise: young people's experiences in dance classes. Journal of Youth Studies, 11 (6), 701-709.

Gentile, B., Grabe, S., Dolan-Pascoe, B., Twenge, J. M., Wells, B. E., & Maitino, A. (2009). Gender differences in domain-specific self-esteem: A meta analysis. Review of General Psychology, 13(1), 34–45.

Genzel, L., Quack, A., Jager, E., Konrad, B., Steiger, A. & Dresler, M. (2012). Complex motor sequence skills profit from sleep. Neuropsychobiology, 66, 237-243.

Gondola, J. C. (1986). The enhancement of creativity through long and short-term exercise programs. Journal of Social Behavior and Personality, 1, 77-82.

Gondola, J. C. (1987). The effects of a single bout of aerobic dancing on selected tests of creativity. Journal of Social Behavior and Personality, 2, 275-278.

Gondola, J. C. & Tuckman, B. W. (1985). Effects of a systematic program of exercise on selected measures of creativity. Perceptual and Motor Skills, 60, 53-54.

Gore, G. (1997) The Beat Goes on: Trance, Dance and Tribalism in Rave Culture. In. (Ed) H. Thomas, Dance in the City, MacMillan Press Ltd.

Hackney, M.E., Kantorovich, S., Levin, R., & Earhart, G.M. (2007). Effects of tango on functional mobility in Parkinson's disease: A preliminary study. Journal of Neurologic Physical Therapy, 31, 173–179.?

Hanggi, J., Keoneke, S., Bezzola, L. & Jancke, L. (2010). Structural neuroplasticity in the sensorimotor network of professional female ballet dancers. Human Brain Mapping, 31, 1196-1206.

Harley, T. (2013). The Psychology of Language: From data to theory. (Fourth Ed.) Psychology Press, UK.

Health Development Agency (2000). Arts for health: A review of good practice in community-based arts projects and initiatives which impact on health and wellbeing. ISBN: 1-84279-016-1. Retrieved from www.hda-online.org.uk

Holdsworth, N. (2013). "Boys don't dance, do they?" Research in Dance Education: The Journal of Applied Theatre and Performance, 18(2), 168-178.

Hopkins, D. R., Murrah, B., Hoeger, W. W. K., & Rhodes, C. (1990). Effect of low-impact aerobic dance on the functional fitness of elderly women. The Gerontologist, 30(2), 189-192.

Huang, L., Galinsky, A. D., Gruenfeld, D. H., & Guillory, L. E. (2011). Powerful postures versus powerful roles: Which is the proximate correlate of thought and behaviour? Psychological Science, 22(1), 95-102.

Hugill, N., Fink, B. & Neave, N. (2010). The role of human body movements in mate selection. Evolutionary Psychology, 8(1), 66-89.

Hui, E., Chui, B. T-K., & Woo, J. (2009). Effects of dance on physical and psychological well-being in older persons. Archives of genrontology and Geriatrics, 49, e45-e50.

Hulme, C., Roodenrys, S., Schweickert, R., Brown, G. D. A., Martin, S. & Stuart, G. (1997). Word-Frequency Effects on Short-Term Memory Tasks: Evidence for a Redintegration Process in Immediate Serial Recall. Journal of Experimental Psychology: Learning, Memory, and Cognition, 23(5), 1217-1232.

Hutchinson Guest, A. (2005). Labanotation: The system for analysing and recording movement. Routledge. New York & London.

James, W. (1890). The Principles of General Psychology (Vol. 1). New York: Holt

Janata, P., Tomic, S. T. & Haberman, J. M. (2012). Sensorimotor coupling in music and the psychology of the groove. Journal of Experimental Psychology: General, 141(1), 54-75.

Jeong, Y-J., Hong, S-C., Lee, M. S., Park, M-C., Kim, Y-K. & SuhC-M. (2005). Dance movement therapy improves emotiponal responses and modulates neurohormones in adolescents with mild depression. International Journal of Neuroscience, 115:12, 1711-1720.

Johansson, G. (1973). Visual perception of biological motion and a model for its analysis. Perception and Psychophysics, 14, 201–211.

Jousilahti, P., Vartiainen, E., Tuomilehto, J., & Puska, P. (1999). Sex, Age, Cardiovascular Risk Factors, and Coronary Heart Disease. Circulation, 99, 1165-1172.

Kaeppler, A. L. (1978). Dance in Anthropological Perspective. Annual Review of Anthropology, 7, 31-49.

Kaltsatou, A. C. H., Kouidi, E. I., Anifanti, M. A., Douka, S. I., & Deligiannis, A. P. (2014). Functional and psychosocial effects of either a traditional dancing or a formal exercising training program in patients with chronic heart failure: a comparative randomized controlled study, Clinical Rehabilitation, 28(2), 128-138.

Kempler, L. & Richmond, J. L. (2012). Effect of sleep on gross motor memory. Memory, 20(8), 907-914.

Kiefer, A. W., Riley, M. A., Shockley, K., Sitton, C. A., Hewett, T. E., Cummins-Sebree, S. & Haas, J. G. (2011). Multi-segmental postural coordination in professional ballet dancers. Gait & Posture, 34, 76-80.

Kim, C-G., June K-J., & Song, R. (2003). Effects of a health-promotion program on cardiovascular risk factors, health behaviors, and life satisfaction in institutionalized elderly women. International Journal of Nursing Studies, 40, 375-381.

Kirsch, L. P., Drommelschmidt, K. A. & Cross, E. S. (2013). The impact of sensorimotor experience on affective evaluation of dance. Frontiers in Human Neuroscience, 7:521.

Kirschner, S. & Tomasello, M. (2009). Joint drumming: Social context facilitates synchronization in preschool children. Journal of Experimental Child Psychology, 102(3), 299-314.

Koch, S. C., Morlinghaus, K. & Fuchs, T. (2007). The joy dance: Specific effects of a single dance intervention on psychiatric patients with depression. The Arts in Psychotherapy, 24, 340-349.

Kuczynski, M., Szymanska, M. & Biec, E. (2011). Dual- task effect on postural control in high-level competitive dancers. Journal of Sports Sciences, 29(5), 539-545.

Laban, R. (1947). Effort. London, England: Macdonald and Evans.

Lane, A. M., Hewston, R., Redding, E., & Whyte, G. P. (2003). Mood changes following modern-dance classes. Social Behavior and Personality 31(5): 453–460.

Lane, A., & Lovejoy, D. J. (2001). The effects of exercise on mood change: The moderating effect of depressed mood. Journal of Sports Medicine and Physical Fitness, 41(4), 539–545.

Leung, A, K.-Y., Kim, S., Polman, E., Ong, L. S., Qiu, L., Goncalo, J. A. & Sanchez-Burks, J. (2012). Embodied metaphors and creative "acts". Psychological Science, 23(5), 502-509.

Lewis, C. (2012). The Relationship between Improvisation and Cognition. Thesis submitted for the degree of Doctor of Philosophy. University of Hertfordshire.

Lewis, C., Annett, L. E., Davenport, S., Hall, A. A. & Lovatt, P. J. (2016). Mood changes following social dance sessions in people with Parkinson's disease. Journal of Health Psychology, 21 (4), 483-492.

Lewis, C. & Lovatt, P. J. (2013). Breaking away from set patterns of thinking: Improvisation and divergent thinking. Thinking Skills and Creativity, 9, 46-58.

Lorenz, K. (1935). Der kumpan in der umvelt des vogels. Der artgenosse als auslosendes moment soialer verhaltensweisen. Journal fur Ornithologie, 83, 137-215, 289-413.

Lovatt, P. J. (2011). Dance confidence, age and gender. Personality and Individual Differences, 50, 668-672.

Madison, G. (2006). Experiencing groove induced by music: Consistency and phenomenology. Music Perception, 24, 201–208.

Mackrell, J. (1997). Reading Dance. Michael Joseph, London, UK.

Mavradis, G., Filippou, F., Rokka, S. & Al, E. (2004). The effect of a health-related aerobic dance program on elementary school children. Journal of Human Movement Studies, 47: 337–349.

McClelland, J. L., McNaughton, B. L. & O'Reilly, R. C. (1995). Why there are complementary learning systems in the hippocampus and neocortex: insights from the successes and failures of connectionist models of learning and memory. Psychological Review, 102(3), 419-457.

Miller, G. A. (1956). The magic number seven, plus or minus two: Some limits on our capacity for processing information. Psychological Review, 63, 81-93.

Montero, B. (2012). Practice makes perfect: the effect of dance training on the aesthetic judge. Phenomenology and the Cognitive Sciences, 11, 59-68.

Nigmatullina, Y., Hellyer, P. J., Nachev, P., David J. Sharp, D. J. & Seemungal, B. M. (2015). The Neuroanatomical Correlates of Training-Related Perceptuo-Reflex Uncoupling in Dancers. Cerebral Cortex, 25(2): 554-562.

REFERENCES

Office for National Statistics (2005). Marriages: Age at marriage by sex and previous marital status, 1991, 2001 and 2003–2005. Population Trends, 127.

Osterhammel, P., Terkildsen, K., & Zilstorff, K. (1968). Vestibular habituation in ballet dancers. Acta Otolaryngol. 66(3):221–228.

Pavlova, M., Krageloh-Mann, I., Sokolov, A. & Birbaumer, N. (2001). Recognition of point-light biological motion displays by young children. Perception, 30, 925-933.

Peelen, M. V., Atkinson, A. P., Andersson, F. & Vuilleumier, P. (2007). Emotional modulation of body-selective visual areas. Social, Cognitive and Affective Neuroscience, 2, 274-283.

Penedo, F.J., & Dahn, J.R., (2005). Exercise and well-being: a review of mental and physical benefits associated with physical activity. Current Opinions in Psyciatry. 18, 189–193.

Perrin, P., Deviterne, D., Hugel, F. & Perrot, C. (2002). Judo, better than dance, develops sensorimotor adaptabilities involved in balance control. Gait & Posture, 15, 187-194.

Phillips-Silver, J. & Trainor, L. J. (2005). Feeling the beat: Movement influences infant rhythm perception. Science, 308, 1430.

Plihal, W. & Born, J. (1997). Effects of early and late nocturnal sleep on declarative and procedural memory. Journal of Cognitive Neuroscience, 9, 534-547.

Provasi, J. & Bobin-Begue, A. (2003). Spontaneous motor tempo and rhythmical synchronisation in 2 1/2- and 4-year-old children. International Journal of Behavioral Development, 27(3), 220-231.

Pullum, G. K. (1991). The Great Eskimo Vocabulary Hoax and other Irreverent Essays on the Study of Language. The University of Chicago Press. Chicago and London.

Quin, E., Frazer, L., & Redding, E., (2007). The health benefits of creative dance: improving children's physical and psychological wellbeing. Education and Health, 25(2), 31-33.

Reed, S. A. (1998). The politics and poetics of dance. Annual Review of Anthropology, 27, 503-532.

Riskin, J. H. & Gotay, C. C. (1982). Physical posture: Could it have regulatory or feedback effects on motivation and emotion? Motivation and Emotion, 6, 273-298.

Risner, D. (2009). What we know about boys who dance: The limits of contemporary masculinity and dance education. In (Eds.) J. Fisher, & A. Shay, When Men Dance: Choreographing masculinities across borders. Oxford University Press.

Robins, R. W., & Trzesniewski, K. H. (2005). Self-esteem development across the lifespan. Current Directions in Psychological Science, 14(3), 158–162.

Rozin, P., Taylor, C., Ross, L., Bennett, G. and Hejmadi, A. (2005). General and specific abilities to recognize negative emotions, especially disgust, as portrayed in the face and the body. Cognition and Emotion, 19 (3), 397-412.

Sakata, M., Shiba, M., Maiya, K. and Tadenuma, M. (2004). Human Body as the medium in dance movement. International Journal of Human-computer Interaction, 17(3), 427-444.

Sforzo, G. A., McManis, B. G., Black, D., Luniewski, D., Scriber, K. C. (1995). Resilience to exercise detraining in healthy older adults. Journal of the American Geriatrics Society, 43:209–215.?

Shallice, T. & Warrington, E. K. (1970). Independent functioning of verbal memory stores: A neuropsychological study. Quarterly Journal of Experimental Psychology, 22, 261-273.

Shanks, D. (2005). Implicit Learning. In (Eds.) K. Lamberts & R. L. Goldstone Handbook of Cognition, SAGE Publications Ltd. UK.

Shepard, R. N. & Metzler, J. (1971). Mental rotation of three-dimensional objects. Science, 171, 701–703.

Smith, C., & MacNeill, C. (1994). Impaired motor memory for a pursuit rotor task following Stage 2 sleep loss in college students. Journal of Sleep Research, 3, 206-213.

South, J. (2006). Community arts for health: an evaluation of a district programme. Health Education, 106 (2), 155-168.

Spalding, D. A. (1873/1954). Instinct: With original observations of young animals. The British Journal of Animal Behaviour, 2(1), 2-11.

Steffensen, M. S., Joag-dev, C. & Anderson, R. C. (1979). A cross-cultural perspective on reading comprehension. Reading Research Quarterly, 15, 10-29.

Steinberg, H., Sykes, E. A., Moss, T., Lowery, S., LeBoutillier, N. & Dewey, A. (1997). Exercise enhances creativity independently of mood. British Journal of Sports Medicine, 31, 240-245.

Tiedens, L. Z. & Fragale, A. R. (2003). Power moves: Complementarity in dominant and submissive nonverbal behaviour. Journal of Personality and Social Psychology, 84, 558-568.

Tomei, A. & Grivel, J. (2014). Body posture and the feeling of social closeness: Exploratory study in a naturalistic setting. Current Psychology, 33, 35-46.

Tse, D., Langston, R. F., Kakeyama, M., Bethus, I., Spooner, P. A., Wood, E. R., Witter, M. P. & Morris, R. G. M. (2007). Schemas and memory consolidation. Science, 316, 76-82.

Tse, D., Takeuchi, T., Kakayama, M., Kajii, Y., Okuno, H., Tohyama, C., Bito, H. & Morris, R. G. M. (2011). Schema-dependent gene activation and memory encoding in neocortex. Science, 333, 891-895.

van Baar, M.E., Dekker, J., Oostendorp, R.A., Bijl, D., Voorn, T.B., Bijlsma, J.W., (2001). Effectiveness of exercise in patients with osteoarthritis of hip of knee: nine months' follow up. Annals of the rheumatic diseases. 60, 1123–1130.

van Kesteren, M. T., Fernandez, G., Norris, D. G. & Hermans, E. J. (2010). Persistent schema-dependent hippocampal-neocortical connectivity during memory encoding and postencoding rest in humans. Proceedings of the National Academy of Science, 107, 7550-7555.

van Kesteren, M. T., Rijpkema, M., Ruiter, D. J. & Fernandez, G., (2010). Retrieval of associative information congruent with prior knowledge is related to increased medial prefrontal activation and connectivity. Journal of Neuroscience, 30, 15888-15894.

Viscki-Stalec, N., Stalec, J., Katic, R., Podvorac, D. & Katovic, D. (2007). The impact of dance-aerobics training on the morpho-motor status in female high-schoolers. Collegium Antropologicum, 31: 259–266.

Vogel, L. (2012). Dance lessons on a personal health budget. Canadian Medical Association Journal, 184(9), E444-E446.

Walk, R. D. & Homan, C. P. (1984). Emotion and dance in dynamic light displays. Bulletin of the Psychonomic Society, 22(5), 437-440.

Walker, M. P. (2005). A refined model of sleep and the time course of memory formation. Behavioural and Brain Sciences, 28, 51-64.

Walker, M. P., Brakefield, T., Morgan, A., Hobson, J. A. & Stickgold, R. (2002). Practice with sleep makes perfect: Sleep-dependent motor skill learning. Neuron, 35, 205-211.

Waller, J. (2009). The Dancing Plague: The strange, true story of an extraordinary illness. Sourcebooks Inc.

Warburton, E. C., Wilson, M., Lynch, M. & Cuykendall, S. (2013). The cognitive benefits of movement reduction: Evidence from dance marking. Psychological Science, 24(9), 1732-1739.

Wessel, T. R., Arant, C. B., Olson, M. B., Johnson, B. D., Reis, S. E., Sharaf, B. L., Shaw, L. J., Handberg, E., Sopko, G., Kelsey, S. F., pepine, C. J., & Merz, C. N. B. (2004). Relationship of physical fitness vs body mass index with coronary artery disease and cardiovascular events in women. Journal of the American Medical Association, 292(10):1179–1187.

White, G., Fishbein, S. & Rutstein, J. (1981). Passionate love and the misattribution of arousal. Journal of Personality and Social Psychology, 41, 56-62.

Whitehead, J. R. (1995). A study of children's physical self-perceptions using an adapted physical self-perception profile question- naire. Pediatric Exercise Science, 7, 132–151.

Winkler, I., Haden, G. P., Ladinig, O., Sziller, I. & Honing, H. (2009). Newborn infants detect the beat in music. Proceedings of the National Academy of Sciences, 106(7), 2468-2471.

Witek, M. A. G., Clarke, E. F., Wallentin, M., Kringelbach, M. L. & Vuust, P. (2014). Syncopation, body-movement and pleasure in groove music. PLoS ONE, 9(4) e94446.

Woodward, K. (1991). Aging and its Discontents: Freud and Other Fictions. Bloomington: Indiana. University Press.

Zentner, M. & Eerola, T. (2010). Rhythmic engagement with music in infancy. Proceedings of the National Academy of Sciences, 107(13), 5768-5773.

Notes